Bioenergy for Power Generation, Transportation and Climate Change Mitigation

Online at: https://doi.org/10.1088/978-0-7503-3555-3

IOP Series in Renewable and Sustainable Power

The IOP Series in Renewable and Sustainable Power aims to bring together topics relating to renewable energy, from generation to transmission, storage, integration, and use patterns, with a particular focus on systems-level and interdisciplinary discussions. It is intended to provide a state-of-the-art resource for all researchers involved in the power conversation.

Series Editor
Professor David Elliott
Open University, UK

About the Editor
David Elliott is emeritus Professor of Technology Policy at the Open University, where he developed courses and research on technological innovation, focusing on renewable energy policy. Since retirement, he has continued to write extensively on that topic, including a series of books for IOP Publishing and a weekly blog post for *Physics World* (physicsworld.com/author/david-elliott)

About the Series
Renewable and sustainable energy systems offer the potential for long-term solutions to the world's growing energy needs, operating at a broad array of scales and technology levels. The IOP Series in Renewable and Sustainable Power aims to bring together topics relating to renewable energy, from generation to transmission, storage, integration, and use patterns, with a particular focus on systems-level and interdisciplinary discussions. It is intended to provide a state-of-the-art resource for all researchers involved in the power conversation.

We welcome proposals in all areas of renewable energy including (but not limited to): wind power, wave power, tidal power, hydroelectric power, PV/solar power, geothermal power, bioenergy, heating, grid balancing and integration, energy storage, energy efficiency, carbon capture, fuel cells, power to gas, electric/green transport, and energy saving and efficiency.

Authors are encouraged to take advantage of electronic publication through the use of colour, animations, video, data files, and interactive elements, all of which provide opportunities to enhance the reader experience.

Do you have an idea for a book you'd like to explore?
We are currently commissioning for the series; if you are interested in writing or editing a book please contact Caroline Mitchell at caroline.mitchell@ioppublishing.org.

Bioenergy for Power Generation, Transportation and Climate Change Mitigation

Anirudh Singh
School of Sciences, University of Southern Queensland, Queensland QLD4300, Australia

IOP Publishing, Bristol, UK

ISBN 978-0-7503-3555-3 (ebook)
ISBN 978-0-7503-3553-9 (print)
ISBN 978-0-7503-3556-0 (myPrint)
ISBN 978-0-7503-3554-6 (mobi)

DOI 10.1088/978-0-7503-3555-3

Version: 20230401

IOP ebooks

British Library Cataloguing-in-Publication Data: A catalogue record for this book is available from the British Library.

Published by IOP Publishing, wholly owned by The Institute of Physics, London

IOP Publishing, No.2 The Distillery, Glassfields, Avon Street, Bristol, BS2 0GR, UK

US Office: IOP Publishing, Inc., 190 North Independence Mall West, Suite 601, Philadelphia, PA 19106, USA

To all those (big and small), who seek to aspire—

The greater part of success is scoping your work.

But humility and trust in people works as well.

Contents

Preface

This book is primarily about the production and use of bioenergy, a form of renewable energy that has been gaining increasing attention recently due to its role in climate change mitigation. It also touches briefly on the human side of energy use.

It was originally meant as a reference text for graduate and academic researchers who wished to keep abreast of the latest development in bioenergy and climate change mitigation. But the growing interest in the search for new fuels and technology for emissions reduction made it necessary to enlarge the scope of the book to include new readership that comprise energy practitioners and decision-makers from a wide range of the governance and economic sectors.

A book in bioenergy and climate change mitigation necessarily involves several disciplines of study (from both the humanities and the sciences). The typical reader must possess a broad understanding of the whole range of these subjects at an adequate level. The book has tried to develop such understanding by treating each area of study from the basics upwards and by taking a first principles approach wherever possible.

The format of the chapters has been chosen to optimise their didactic value. The introduction to each chapter contains a synopsis of the content that is to follow. Material relevant but not necessary to the main theme is included in boxes to enable the smooth development of the chapter. Case studies have been included wherever possible, to relate the academic content to the real world context. Chapters and sections have been cross-referenced liberally in the text to assist the reader's recall. Questions and exercises have been included, and the reader is encouraged to at least peruse these whenever possible.

The first three parts are more technical in nature than the last. They assume a year 12 knowledge of chemistry and physics. For those who are daunted by the equations, the text provides sufficient information for their needs.

The book is divided into four parts.

Part I shows how bioenergy is produced in its various forms and converted to more useful forms. It begins with an overview of bioenergy at the global level. It then looks at solid biomass as an energy source, and describes the thermochemical, biochemical and electrochemical pathways by which it is converted to forms more amenable for use.

Part II introduces the principles and techniques of generating and storing electrical energy from bioenergy. This part begins by introducing the principles of thermodynamics as they apply to the operation of heat engines, and those employing the Rankine and Brayton cycles in particular. It goes on to describe the principles of operations of thermal power plants, and provides examples of the range of bioenergy-fired power plants in industrial application. It ends by looking at the roles of electrolysers, fuel cells and batteries in the storage and conversion of electrochemical energy, and traces the evolution of primary and rechargeable batteries in use today.

Part III discusses the science, technology and legislation behind the production and use of liquid and gaseous biofuels. It introduces the production and use of bioethanol, biobutanol, biodiesel, biogas and biomethane, as well as the emerging biofuels HVO, syngas, pyrolysis oil and hydrogen.

Part IV is about the human dimension of energy. It shows how global politics and policy can act as both drivers and threats to renewable energy production and climate change mitigation. This final part of the book demonstrates how global energy trade and the need for climate change mitigation and sustainable development influence global green energy production. It also shows how climate change policy imperatives are transforming energy technology and infrastructure on a global scale. It examines the current policies for addressing global challenges, and argues that an integrated approach, involving a single framework, is needed to address them adequately.

Parts I–III provide material that is well-suited as a reference text for graduate and academic researchers working in the various areas of bioenergy research and development. They deal with the basic science and technology of the production and use of solid, liquid and gaseous bioenergy. Part IV switches attention to energy in the human context, and is accessible to the widest readership, including those in commerce, industry and politics who are interested in the socio-economic and climate impacts of energy supply on a global scale. This part is of special relevance to those working in the areas of climate change mitigation and the current energy transition, and provides a repository of information on recent developments to inform academics, strategists as well as activists in these areas.

Anirudh Singh
Melbourne
December 2022

Acknowledgements

Writing this book was not easy. It began as a simple (but much-needed) reference text in the area of bioenergy for graduate researchers and their supervisors, but expanded in its scope as the subject matter gained increasing relevance to current developments in the energy field. I am therefore deeply indebted to the many people who helped in their own way in making it possible.

Firstly my grateful thanks to the following academic colleagues from the Australian National University who provided direct support relating to the ANU case study:

- Professor Mark Howden—for enabling the gathering of information on ANU's Grand Challenge program;
- Juliette Blaszczyk and Nicole Foronda—for assisting in establishing contact with the Grand Challenge team members;
- Dr Fiona Beck—for the several virtual meetings I had with her, and for providing the comprehensive introduction to the Grand Challenge project as well as reviewing the final output of the case study;
- Dr Igor Skryabin—for the virtual meeting we had, and the insights he provided on ANU's work on the energy transition;
- Professor Asmi Wood of the ANU Law School—for helpful clarifications on indigenous people and land rights.

I would like to thank Professor Dewei Chu, leader of UNSW's Nanoionic Materials Group, for informative discussions and copies of their publications on moisture electric generators (MEGs) published recently in two journals.

My appreciation to Bijma and Asha for coming to my rescue by taking up the burden of drawing the figures for the chapters, and also to my whole family and friends for their patience and forbearance while I worked on the book at the expense of the quality time I would otherwise have spent with them.

Lastly, my special thanks to Caroline Mitchell, for her profound understanding and her sincere words of encouragement during the (longer than expected) period of writing the book.

Author biography

Anirudh Singh

Anirudh Singh, PhD, is a researcher/writer/consultant in Renewable Energy and Climate Change Mitigation and a former Professor in Renewable Energy. He is currently an Adjunct Professor at the University of Southern Queensland.

During his time at The University of Fiji (2017–21), he developed a Masters in Renewable Energy Management program and wrote and edited books in renewable energy and climate change mitigation (for IOP Publishing and Springer Nature, respectively). He also authored an expository book in physics titled *Concepts and the Foundations of Physics* for AIP Publishing.

Earlier in his academic career at The University of the South Pacific (2006–16), Anirudh was project leader for two EU-funded energy capacity-building projects involving a global consortium of universities. He was also a Section Editor for the *Handbook of Climate Change Adaptation* (Springer) and Expert Reviewer of the Second and First Order Drafts of the Working Group II contribution to the *IPCC Fifth Assessment Report* (AR5).

Anirudh earned his PhD in Condensed Matter Physics from Leicester University, England, and an MSc in Theoretical Nuclear Physics from Auckland University, New Zealand. Following his PhD studies he worked as a researcher in muon implantation studies using the facilities at the Rutherford Appleton Laboratory in the UK and the TRIUMF facility in Canada.

Anirudh's current interest is following and assisting in the development of the ongoing energy transition towards achieving the goal of restricting global temperature rise to 1.5 °C. He is also keenly interested in advocating for an integrated approach (based on a single policy framework) to addressing the global challenges of poverty and economic inequity, pandemics, climate change and natural disasters that the world is faced with today.

linkedin.com/in/anirudh-singh-aa984416

Part I

Introduction to bioenergy

Chapter 1

Renewable energy and biomass

This introductory chapter lays down the groundwork for discussions on the role of bioenergy for power generation, transportation and climate change mitigation that provides the focus of interest for the rest of the book. It begins by introducing the language used to describe how energy is used by the various energy use sectors of current day economies. The nomenclature used by the energy industry to describe the various stages of energy production and energy flows is established. The chapter then puts bioenergy into a global perspective by considering its relative importance to the global final energy consumption as a whole, and renewable energy production in particular.

1.1 Introduction

Energy is a common everyday commodity and an essential for life, forming an indispensable ingredient of all sectors of economies and affecting all spheres of life. It is used in various forms and for a variety of purposes, ranging from a country's domestic and industrial needs and extending to its operational and defence requirements.

The crucial role energy plays in the global economy was demonstrated graphically recently with the onset of the COVID-19 pandemic. To contain the spread of this highly contagious virus, nations were forced into taking drastic confinement measures such as lockdowns and national border closures. An unintentional but inevitable effect of such measures was the slowing down of economic activities, much of which depend on transportation, which in turn requires energy for its sustainability. The tourism sector was hit the hardest, with the global airline industry being virtually brought to a standstill. What followed inevitably was a sharp fall in the global fuel demand for the transportation and manufacturing sectors. This was a stark reminder of the link between the energy supply and the economy of a country.

Most of the energy used in the world today is fossil fuel energy, and most of it is used for transportation and power generation. However, fossil fuel use has its

doi:10.1088/978-0-7503-3555-3ch1

disadvantages, the most obvious being its impact on climate due to the release of carbon dioxide. The other is the non-ubiquitous nature of its availability. Not all countries of the world have their own oil resources, but they all need to access this energy resource to fuel their manufacturing, transportation and power generation sectors. The resulting supply chain issues lead to a differential pricing structure where the remotest countries bear the highest costs. Unfortunately, many of these nations are amongst the poorest of the world.

For these and other reasons, there has been a global movement towards alternative forms of energy, especially for transportation and power generation. Amongst these alternative forms, bioenergy (in the forms of solid biomass and liquid and gaseous biofuels) stands out as an obvious choice. In the first instance, these energy alternatives are renewable and thus less greenhouse gas emitting than their fossil fuel counterparts. They also have the advantage that they are relatively more accessible than other fuels in countries lacking indigenous fossil fuel resources of their own.

The share of these forms of energy in the global energy mix, however, remains low. There are several reasons for this, the availability of suitable technology for their production and use being perhaps the most important [1].

The principal aim of this book is to consider the various forms of bioenergy as clean energy alternatives for power generation, transportation and climate change mitigation. It considers all aspects of the production and use of these fuel sources. But before one ventures into the details of these aspects of bioenergy, it is illustrative to consider the global status of bioenergy production and use.

This chapter provides such an overview. It begins in section 1.2 by introducing the various energy sectors of an economy, and provides the required familiarity with the language and nomenclature used to describe energy production and use. Section 1.3 describes the energy flows in a country between its primary energy sources and the country's end-use energy. Section 1.4 provides a perspective on the global primary energy supply and the total final energy use. The last three sections elaborate on the share of renewable energy in the global final energy mix, and the proportion of this form of energy that is provided by bioenergy.

1.2 The energy use sectors

We need to categorise energy use to understand it better. This is best done by asking the questions

- where does energy come from, and
- what is energy used for?

The most-used forms of energy today are oil, coal, natural gas, nuclear energy, hydro, wind, solar, solid biomass, liquid and gaseous biofuels, geothermal energy, and forms of ocean energy to a lesser extent. These may be broadly characterised according to their sources of origin into:

- **fossil fuels** (oil, coal, natural gas) which are mineral fuels that occur underground;

- **nuclear energy** which is also derived from underground sources; and
- **renewable energy** (hydro, wind, solar, biomass, biofuels, ocean energy), all of which are available from terrestrial sources and are readily replenished.

Geothermal energy is a bit of an oddity. Although it originates from underground sources and is strictly non-renewable, it is always classified as renewable energy. It is customary to group fossil fuels and nuclear energy into the **non-renewables**, and solar, hydro, wind, biomass, biofuel, geothermal and ocean energy into the **renewables**. It must be noted, however, that perhaps the most common form of energy used in everyday life is electricity. This energy is a derived form of energy and will be discussed at length later in this chapter.

Energy is used for an enormous variety of reasons by people and nations as a whole. This usage is usually divided into **energy use sectors**. A commonly used classification for these economic sectors consists of the **domestic**, **commercial**, **industrial**, **transport** and electrical **power generation** sectors. They may sometimes also include the **mining** and **agriculture** sectors. But there are other sectorial classifications possible also. For instance, another system categorises them into the power generation, transport, heating and cooling sectors. Yet others may divide them into building, industry, transport and other sectors.

Table 1.1 gives an example of how electrical energy is used by the economic sectors in developed and developing countries. The similarity is striking.

Table 1.1. Comparison of electrical energy use by the various economic sectors of developed and developing countries. Adapted from [2].

Sector	Electricity use (USA) (%)	Electricity use (Fiji) (%)
Residential	37	30
Commercial	35	45
Industrial	27	25
Transportation	0.2	0.0
Total	**99.2**	**100**

1.3 Energy flow—from primary to end-use energy

We saw above that electricity is a form of energy that is derived from other forms of energy. Clearly there must be transformations that take place between the various forms of energy. An example is provided by electricity generation at a diesel power station. The diesel fuel is needed to produce electricity. However, this fuel is in turn produced from crude oil at an oil refinery. We therefore have an energy chain where the energy in the original medium (crude oil) is transferred to the energy contained in diesel and eventually transformed into electrical energy in the power plant.

Electricity is usually the final form of energy required by the end user in this energy chain. It is appropriately called **end-use energy** (or sometimes **final energy**). Crude oil, the form of energy at the start of this chain, is called **primary energy**.

All the energy used in a country originates from primary sources of energy. Typically, these forms of energy undergo various transformations before they are made available as end-use energy to the consumer. There are, however, exceptions where the energy is used directly in the form that it entered the country. Examples include LPG or other forms of cooking gas.

To understand the manner in which energy is produced and used in a country in any one year, it is best to picture energy as being produced from an original source, namely the primary energy, and undergoing energy transformations in an energy chain that ends in the final, or end-use energy desired by the consumer. A diagram depicting this process is called the **energy flow chart** of the country. Figure 1.1 provides a schematic representation of the process.

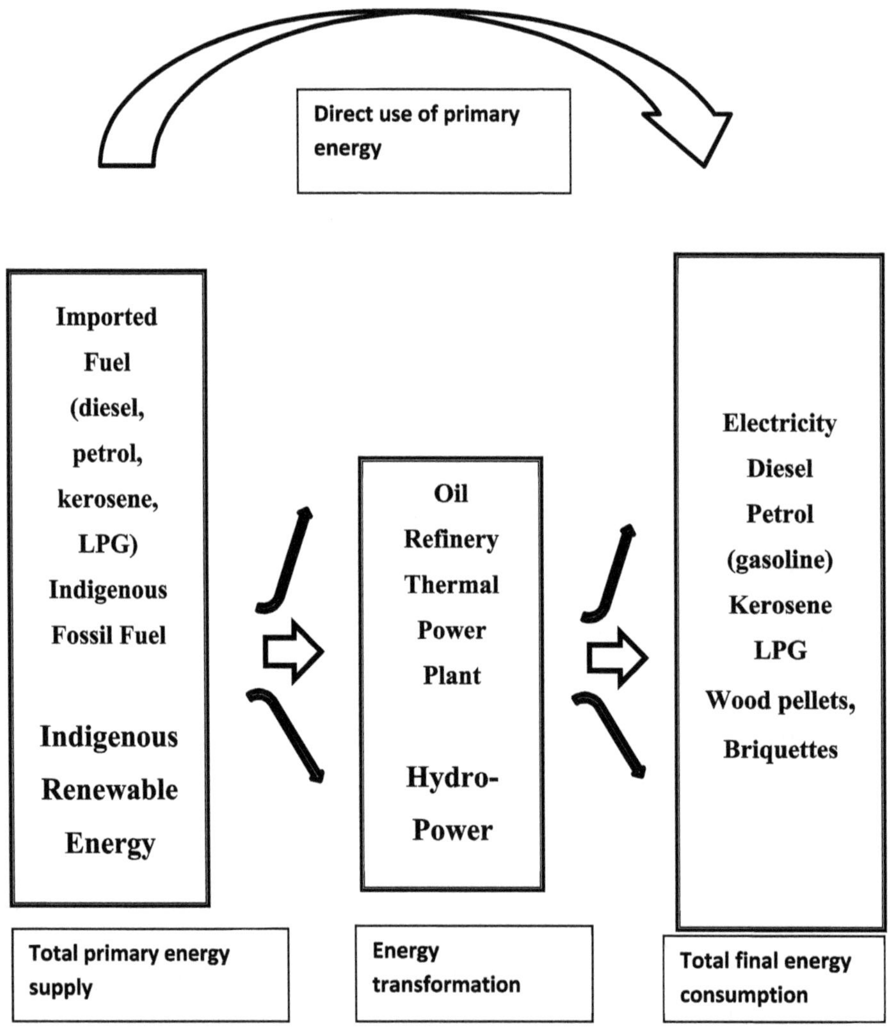

Figure 1.1. A schematic representation of the energy flow chart of a country. Adapted from [2]. Copyright © 2018 Morgan & Claypool Publishers. All rights reserved.

In this figure, primary energy is broadly used to mean all energy that enters the country through importation as well as through the primary energy recourses that are indigenous to the country. Any energy conversion (from primary to end-use energy) takes place in the facilities shown in the central block. Thus, oil refineries may be used in the country to refine imported fossil fuels, hydropower stations convert the potential energy of water in dams and reservoirs to electricity, while wind turbines produce the same output for the wind energy resource available at their specific locations.

The sum total of the energy that enters this chain is called the **Total Primary Energy Supply (TPES).** The sum total of energy available to the user is called the **Total Final Energy Consumption (TFEC)**. As energy losses occur at every conversion, the total final energy is always lower than the total primary energy that is input into this process.

(See exercise 1.1—An energy flow chart of your country.)

1.4 Global primary energy and total final energy consumption

Most energy on the Earth originates from either solar energy or nuclear energy. The exceptions are geothermal and tidal energy, both of which are related to the history of the formation of the solar system.

The total incident solar energy per annum on Earth is 3.4×10^6 EJ [3]. Of this, the amount transformed into biomass via photosynthesis is only 3000 EJ (or 0.078% of total incident radiation). Thus, the photosynthetic capture of solar energy for conversion into biomass energy is a highly inefficient process.

In 2016, the global total primary energy supply (TPES) was 576 EJ [4]. In the same year, the total final energy consumption (TFEC) was only ~400 EJ [4]. This demonstrates clearly that there are always energy losses involved in the conversion of primary energy into the final energy in any country.

Table 1.2 shows the total global primary energy consumption by country in 2018 (note: 1 tonne oil equivalent = 41 868 MJ).

Table 1.2. Global primary energy consumption by country in 2018 (data source: [5]).

	Primary energy consumption	
Region	(Mtoe)	% of total
North America	2832.0	20.4
South and Central America	702.0	5.1
Europe	2050.7	14.8
CIS	930.5	6.7
Middle East	902.3	6.5
Africa	461.5	3.3
Asia Pacific	5985.3	43.2
Total	**13 864.0**	**99.9**

Table 1.3. Global primary energy consumption by sector in 2016 (data source: [4]).

Sector	Global share of primary energy (%)
Building	31
Industry	29
Transport	29
Other	12
Total	**101**

It is interesting to see how primary energy is used by the economic sectors of the world.

Table 1.3 shows how the total primary energy production was consumed by the building, industry, transport and other sectors of the global economy in 2016 [4].

(See exercise 1.2—Converting global primary energy data into absolute units.)

1.5 Share of renewable energy in the global final energy consumption

Having obtained a broad perspective of the total energy production of the world, we now consider how renewable energy (RE) features in this global energy mix. Two important parameters are the share of RE in the total final energy consumption and electricity generation, respectively. We consider the first in this section and the second in the following section.

In 2017, the non-renewables made up 81.9% of the total final energy consumption of the world, with the renewables making up the remainder. These contributed to the total according to the categories shown in table 1.4.

Note that the renewables in the above table have been divided into traditional biomass and the modern renewables. The first category consists of all forms of primary biomass energy that are used traditionally by the poor of the world, and include firewood, crop residues and animal manure. The modern renewables comprise those forms of RE that utilise modern technology in either processing or extracting energy from primary renewable energy sources.

What is notable about RE is that:

- this form of energy still forms only a small fraction of the total final energy consumption of the world, and
- traditional biomass forms a large component of RE.

In recent years, the global percentage share of renewables has been rising rapidly. This is due to a concerted global drive towards reductions in carbon emissions resulting from climate change negotiations, and the rapid reduction in the prices of solar and wind energy.

Table 1.4. Estimated global renewable and non-renewable final energy consumption (%) in 2017 (data source: [4, figure 1, p 31]).

	% Share	Total of category
Non-renewables:		
Fossil fuel	79.7%	
Nuclear	2.2%	
Total non-renewable	**81.9%**	**81.9%**
Traditional biomass	**7.5%**	**7.5%**
Modern renewables:		
Biomass/solar/geothermal heat	4.2%	
Hydro	3.6%	
Wind/solar/biomass/geothermal/ocean Power	2.0%	
Biofuel for transport	1.0%	
Total modern renewables	**10.6%**	**10.6%**
Total final energy consumption		**100.0%**

1.6 Share of renewable energy in global electrical power generation

The last section considered the fraction of the total final energy consumption of the world provided by renewable energy. Electricity plays a dominant role as a form of energy used in the final energy mix, and it is interesting to see what share of this is provided by renewable energy.

The non-renewables play a major role in global electricity generation, providing 74% of the total share in 2018. Amongst the renewables, hydro provides the main energy source. The estimated share of the renewables in the global electrical power generation mix is shown in table 1.5.

Table 1.5. Estimated share of RE in global electricity production in 2018 (data source: [4, figure 8, p 41]).

Non-renewable (fossil and nuclear)	73.8%
Renewable:	
Hydropower 15.8%	
Wind 5.5%	
Solar PV 2.4%	
Biopower 2.2%	
Geothermal, CPS	
Ocean power 0.4%	
Total renewables 26.3%	26.3%
Total	**100%**

(See exercise 1.3—Converting power into absolute units.)

1.7 Global share of biomass energy

How significant is biomass energy on the global scene? We have seen that traditional biomass provides for 7.5% of the total final energy consumption. Table 1.6 reveals that modern use of biomass energy provides for another 5% of the total energy consumption. The table shows how this modern use is distributed amongst the various energy use sectors.

Table 1.6. Estimated share of bioenergy in total final energy consumption (data source: [4, figure 18, p 72]).

Non-biomass	87.6%
Traditional biomass 7.4%	7.4%
Modern bioenergy:	
Heat in industry 2.2%	
Heat in buildings 1.4%	
Transport 1.0%	
Electricity 0.4%	
Total modern bioenergy 5.0%	5.0%
Total	**100%**

As we saw in table 1.5, biomass provided 2.2% of global power production in 2018. Table 1.7 lists some of the important players in global biopower production.

Table 1.7. Bioelectricity generation by country in 2018 (data source: [4, p 73]).

Biopower production (GW) by country		
Year	2017	2018
Biopower generation (GW) by country:		
EU	39	42
UK		7.7
China		17.8
India		10.2
Japan		4
USA		16
Brazil		14.7
World total biopower generation	**121**	**130**

Exercises

Exercise 1.1. An energy flow chart of your country.

Produce an energy flow chart of your country that includes detailed examples of primary energy sources, transformation processes and forms of end-use energy.

Exercise 1.2. Converting global primary energy data into absolute units.

Use the data available to convert % shares in table 1.3 into absolute energy units (EJ).

Exercise 1.3. Converting power into absolute units.

The total power generated from renewable energy in 2018 was 2378 GW [4, p 18]. Convert the % shares in table 1.5 into absolute power units (GW).

References

[1] Singh A and Bijay P 2020 Mitigating through renewable energy: an overview of the requirements and challenges ed A Singh and R Deo *Translating the Paris Agreement into Action in the Pacific* (Cham: Springer Nature) ch 2 pp 29–58

[2] Singh A 2018 *Talking Renewables: A Renewable Energy Primer for Everyone* (Bristol: IOP Publishing)

[3] World Energy Council 2013 *World Energy Resources: Solar* [accessed 5 April 2021]. Available from https://worldenergy.org/assets/images/imported/2013/10/WER_2013_8_Solar_revised.pdf

[4] REN21 2019 *Renewables 2019 Global Status Report 2019* [accessed 5 April 2021]. Available from https://ren21.net/wp-content/uploads/2019/05/gsr_2019_full_report_en.pdf

[5] BP 2019 *BP Statistical Review of World Energy 2019* [accessed 5 April 2021]. Available from https://bp.com/content/dam/bp/business-sites/en/global/corporate/pdfs/energy-economics/statistical-review/bp-stats-review-2019-full-report.pdf

IOP Publishing

Bioenergy for Power Generation, Transportation and Climate Change Mitigation

Anirudh Singh

Chapter 2

Introduction to solid biomass—properties and use

Solid biomass is perhaps the longest known source of bioenergy to humankind. This chapter discusses the various physical forms in which solid biomass is available, how it obtains its energy and stores it chemically, and how this energy is extracted through exothermic chemical reactions. It also shows how the various physical properties of solid biomass are related to its energy content and to each other.

The chapter begins by introducing the basic chemistry of energy storage and retrieval from chemical bonds, and the plant structure and biochemistry required to understand chemical energy storage in ligno-cellulosic materials. After an overview of the forms in which solid biomass is available as an energy source, the important properties of energy content, moisture content and density of solid biomass are introduced. Operational mathematical relationships between these parameters are then provided. The chapter ends with a brief look at the pre-treatment and densification processes that are available to improve the energy content and storage and handling of solid biomass.

2.1 Introduction

Before using biomass as an energy source, it is clearly important to understand how energy is stored in this medium. This chapter examines the basic science of energy in solid biomass. It considers how energy is stored in biomass, and how this energy depends on other biomass properties such as moisture content and density. It also shows how the quality of solid biomass as an energy source may be improved through processes such as pre-treatment.

The chapter begins by showing how biomass obtains its energy from sunlight through the process of photosynthesis, and proceeds to show how energy is stored in the chemical bonds of molecules and released through exothermic reactions such as

doi:10.1088/978-0-7503-3555-3ch2

combustion. It then provides a brief introduction to plant anatomy and biochemistry to reveal where ligno-cellulosic compounds (the energy-rich molecules of interest) are stored in plant cells.

Section 2.3 provides an overview of the sources of solid biomass that are used for power generation and other applications. These sources include energy crops, forestry residues, domestic and animal waste and sewage. The next section considers how solid biomass is used as an energy source by the communities and the sectors of a country's economy.

Section 2.5 introduces the physics of bioenergy use. It begins with the physical concept of energy content and its physical determination using the bomb calorimeter. The energy content is dependent on the moisture content of the biomass, which also determines its density. The next two sections review the mathematical equations describing the energy–moisture and density–moisture relations of solid biomass.

Lastly, section 2.8 of the chapter considers physical and thermal methods of pretreatment and densification of biomass.

2.2 Energy in biomass

Energy is stored in the molecules of biomass through the process of photosynthesis, and can be extracted via the chemical reaction known as oxidation or **combustion**. The following sub-sections provide a simplified overview of these processes, as well as assessing the performance of biomass as an energy source.

2.2.1 Obtaining energy from biomass

How does biomass obtain its energy, and how is it extracted for use?

Biomass is produced when radiant energy from the Sun is converted to the chemical energy stored in glucose molecules synthesised in the leaves of plants through the well-known process of **photosynthesis**. The overall photosynthetic process may be seen as the light-driven conversion of carbon dioxide and water from the atmosphere and the soil into glucose molecules and oxygen, and represented as shown in equation (2.1).

$$6CO_2 + 6H_2O + \text{Light} \rightarrow C_6H_{12}O_6 + 6O_2 \tag{2.1}$$

The actual process is, however, quite complex, involving numerous **metabolic pathways** (chains of reactions that eventually produce the end product such as the glucose molecule) with associated **enzymes** (organic catalysts that make reactions possible at the low temperatures inside organic matter). The reader will become more familiar with the biochemistry of organic synthesis in later chapters. For the present purpose, it will be ample to note that the metabolic pathways consist of [1]:

- an initial reaction, called the **light reaction**, which first converts water from the environment and the molecules adenosine diphosphate (ADP), nicotinamide adenine diphosphate (NADP) and inorganic phosphate molecules already present in the plant into oxygen, protons and the energy-rich molecules adenosine triphosphate (ATP) and NADPH (also called 'reduced NADP'); followed by

- a subsequent reaction, called the **dark reaction**, which then uses carbon dioxide from the atmosphere together with the protons, the ATP and NADPH produced above to produce glucose, while re-generating the other plant molecules mentioned above.

Plants contain a host of other molecules (formed by their own metabolic pathways) in addition to those listed above. All of these have stored energy that was originally supplied by the photosynthesis processes described above.

The energy stored in a molecule is associated with the **chemical bonds** between the atoms of the molecule (the explanation requires knowledge of quantum physics, and will not be considered here). This energy can be 'extracted' by converting this molecule to another molecule (or molecules) with a different bonding structure through a **chemical reaction**. The energy that is produced arises from the difference in bonding structures between the original and new molecules (i.e. the **reactants** and the **products**).

Chemical reactions are called **endothermic** if they require net input of energy to proceed, and **exothermic** if they produce net energy output as a result of the reaction. The chemical energy stored within molecules is released in the form of heat when they undergo exothermic chemical reactions such as combustion, where the molecule reacts with oxygen or some other oxidising agent. It is this heat of combustion that is the useful energy available for application.

Chemists think of the heat energy of combustible compounds by assigning the compound a definite heat content called its **enthalpy** H. The changes in energy that occur during chemical reactions are then measured as **enthalpy changes** ΔH.

As noted above, energy stored in molecules is associated with the chemical bonds between their atoms, and thus enthalpy changes are expected to occur during chemical reactions. For a reaction to proceed, bonds within the starting molecules (the reactants) must be broken, and new bonds formed between the molecular fragments to yield new molecules (the products). The enthalpy change of a chemical reaction is the energy difference between the bond-breaking and bond-making processes.

Consider the combustion reaction

$$CH_4 + 2O_2 \rightarrow CO_2 + H_2O \tag{2.2}$$

in which the reactants methane and oxygen produce the products carbon dioxide and water. For the reaction to proceed, the C–H and the O–O bonds must first be broken, and new C–O and O–H bonds must be formed. The bond-breaking requires energy, and is therefore an endothermic process. The new bonds are more stable than the old ones (i.e. are at a lower **energy level** than the old bonds). The result is that there is net energy released, and the reaction is therefore an exothermic reaction. Figure 2.1 puts all these notions together in the form of an energy level diagram.

These energies of reactions are also described in terms of a diagram called a **reaction profile** (also sometimes called an **activation energy diagram**), (see e.g. [2]).

This shows the activation energy required to elevate the reaction mixture to a transitional state (called the activated complex) and its subsequent transition to the

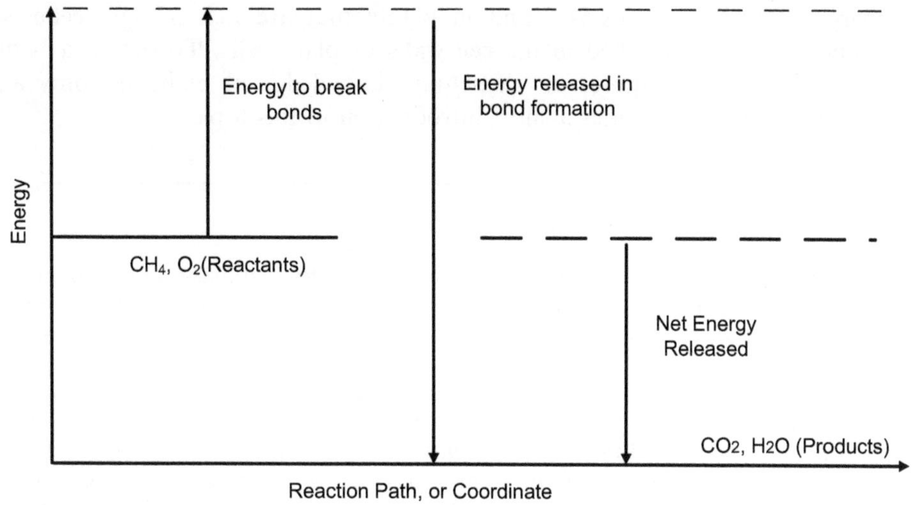

Figure 2.1. Relative energies of the bond-breaking and bond-making processes in an exothermic reaction.

lower energy state of the products. Figure 2.2 shows the essential features of these processes diagrammatically.

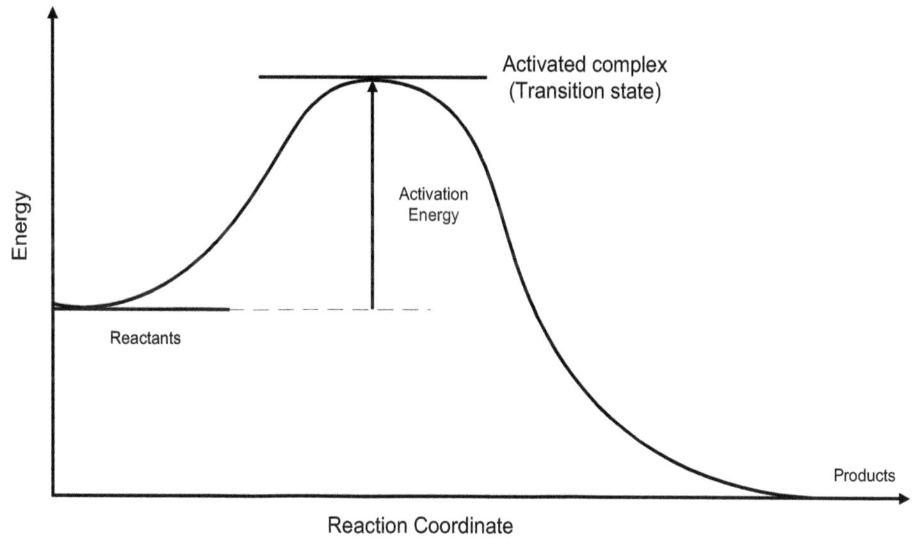

Figure 2.2. The reaction profile (activation energy diagram) of a chemical reaction (see [2]).

One may understand this diagram by equating the bond-breaking energy of the previous diagram to the activation energy, and the net energy released in product formation to the difference between the energy levels of the products and reactants.

2.2.2 Where does the energy in biomass reside?

Plant biomass may be broadly classified as **woody** or **herbaceous**. The natural question to ask is: where is energy stored within the bulk of this biomass? Most of the energy of interest to us is found in plants that are rich in **ligno-cellulosic** compounds. This is contained in the **cell walls** of plant cells. To obtain a better picture of this source requires an elementary knowledge of plant anatomy and biochemistry. Box 2.1 provides a short introduction to this topic.

Box 2.1. Some plant anatomy and biochemistry.

Plants and animals are made up of **cells** arranged in **tissues** that make up their **organs** [3]. The plant organ of main interest to us is the **plant stem**. Its function is to provide structural and functional support to the other organs, namely the leaves, flowers and the roots.

The cross-section of the typical woody stem has **dermal tissues** on the outside that provide protection, **vascular tissues** in the middle that transport water, food and minerals, and **ground tissues** at the centre (in the form of **pith**) and the **cortex**. These tissues are made up of different types of cells, including parenchyma, collenchyma, sclerenchyma and fibre cells, as well as companion cells and sieve tubes.

Plant cells are different from animal cells in that they have **cell walls** as well as structures called **chloroplasts** within them. Figure 2.3 shows the main features of the typical plant cell, including the cell wall and the important sub-cellular units known collectively as **organelles**.

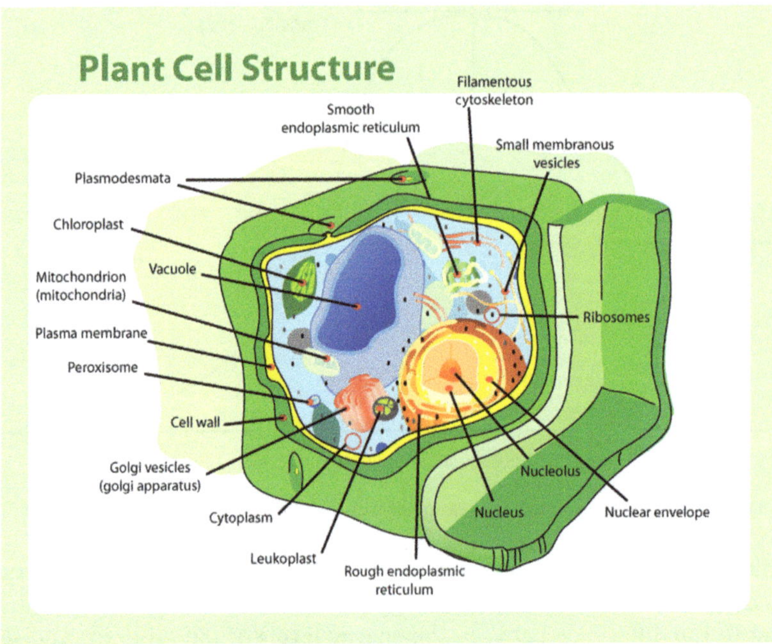

Figure 2.3. Schematic diagram of a plant cell [4]. This plant cell structure image has been obtained by the author from the Wikimedia website, where it is stated to have been released into the public domain. It is included within this article on that basis.

The energy-rich materials contained in plants that are of interest to us are the **ligno-cellulosic compounds**. These molecules are found in the cell walls of the plant cell. They are actually giant macro-molecules, consisting of **polymers** made up of repeating units called **monomers**. The polymers include **cellulose**, **hemicellulose** and **lignin** in the cell walls, as well as **pectin** in the cell lamellae (spaces between cells).

In the plant cell walls, the cellulose, hemicellulose and lignin together form long chains of tube-like structures called **microfibrils**. These are bundled together into **macro-fibrils** which strengthen the walls and provide structural support for the whole plant (for a graphic depiction of these arrangements of fibres, see [5]).

The growing part of a cell wall is called the **primary cell wall**, and consists of cellulose microfibrils linked together with structures containing hemicellulose and pectin. Figure 2.4 illustrates the main features of a segment of the plant cell wall.

The monomer units that make up the ligno-cellulosic polymers are mainly the 6-carbon sugar **glucose** in the case of cellulose, and various 5- and 6-carbon sugars (**pentoses** and **hexoses)** for hemicellulose.

Figures 2.5(a) and (b) show a glucose molecule and a section of the cellulose molecular chain, respectively.

A cellulose molecule consists of 7000–15 000 glucose monomers linked together in a straight chain. Because it is made up of repeated sugar units, cellulose is called a **polysaccharide**. The cellulose polymer is a straight-chain molecule and occurs in a crystalline form.

In contrast to cellulose, hemicellulose consists of a variety of polymers made from pentose and/or hexose monomer units or their acid derivatives. The most common 5-carbon sugar unit is **xylose**. This produces the polymer called **xylan**. Hemicellulose forms an amorphous branched-chain polymer that links the other polymers in the cell wall.

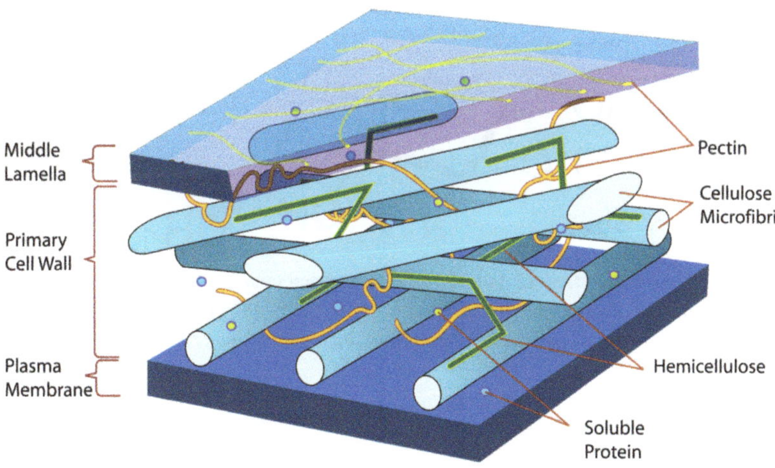

Figure 2.4. Section of a (primary) cell wall, showing cellulose microfibrils interspersed with hemicellulose and pectin structures [6]. This plant cell wall diagram image has been obtained by the author from the Wikimedia website, where it is stated to have been released into the public domain. It is included within this article on that basis.

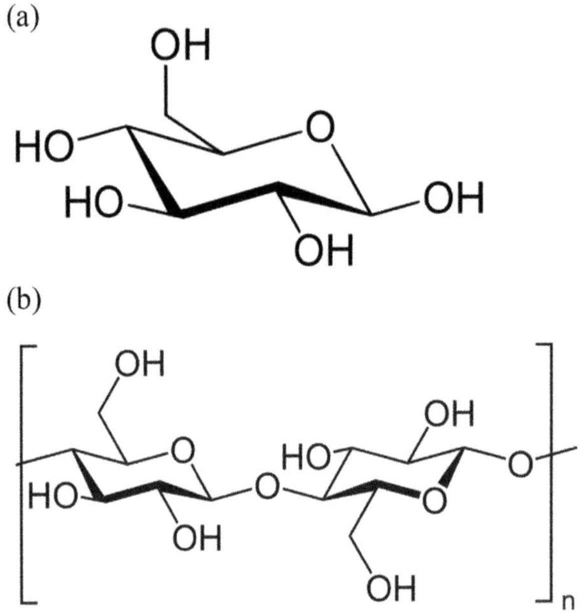

Figure 2.5. (a) A glucose molecule [7]. This beta-D-glucose image has been obtained by the author from the Wikimedia website, where it is stated to have been released into the public domain. It is included within this article on that basis. (b) A representative portion of a cellulose polymer chain, showing the β(1→4) linkage between D-glucose units [8]. This cellulose sessel image has been obtained by the author from the Wikimedia website, where it is stated to have been released into the public domain. It is included within this article on that basis.

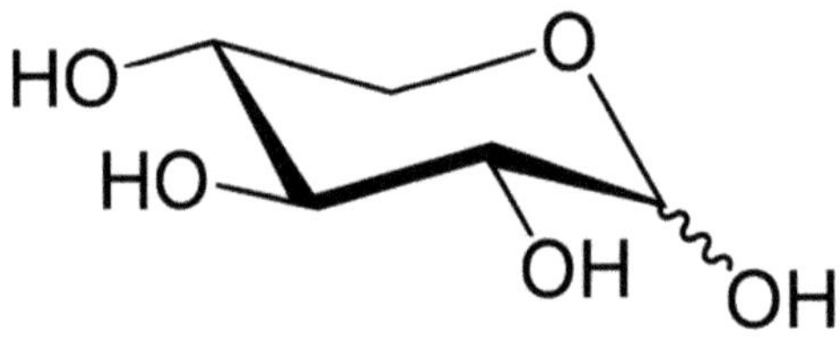

Figure 2.6. The structure of xylose that provides the monomer unit for the form of hemicellulose known as xylan [9]. This D-xylose image has been obtained by the author from the Wikimedia website, where it is stated to have been released into the public domain. It is included within this article on that basis.

Figure 2.6 shows a single molecule of the pentose sugar, xylose, which provides the basic monomer unit for the formation of xylan.

Lignin polymers are made up of units called **monolignols**, the three most common being the **phenolic alcohols (paracoumaryl alcohol**, **coniferyl alcohol** and **sinapyl alcohol**).

1
2
3

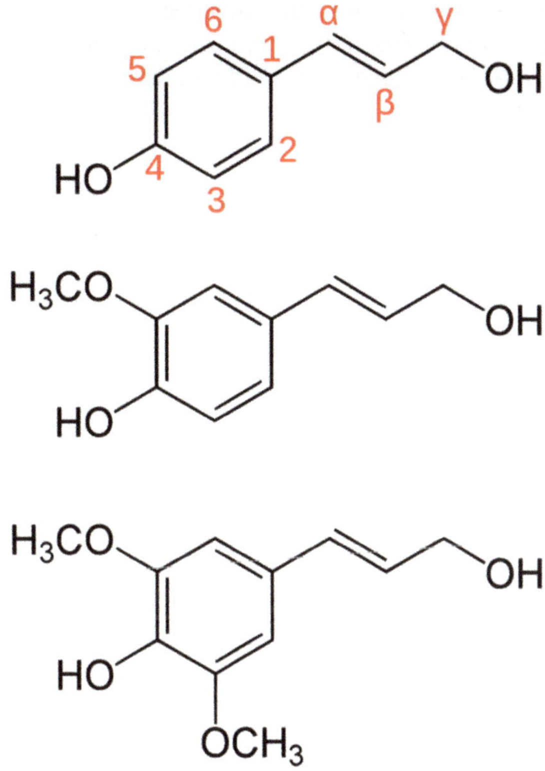

Figure 2.7. The three phenolic alcohols that provide the monomers for lignin: (1) *paracoumaryl alcohol*, (2) *coniferyl alcohol* and (3) *sinapyl alcohol* [10]. This monolignols general image has been obtained by the author from the Wikimedia website, where it is stated to have been released into the public domain. It is included within this article on that basis.

Figure 2.7 shows the structures of the three most common monolignols providing the building blocks for lignin.

(See exercise 2.1—Anatomy of woody plants.)

2.2.3 Biomass as an energy source

Biomass is an easily available source of energy, and is renewable. But how good is its performance as an energy source?

Compared to other sources of energy, biomass has two disadvantages. Firstly, its energy content is relatively lower. (The energy content of biomass is defined fully in section 2.4.) Secondly, it has a high moisture content which lowers its energy content and generally affects its performance as an energy fuel. The lowering of the energy content is understood most readily by realising that energy is extracted from biomass as the heat of combustion when the biomass is burnt. As water does not burn, its presence in the biomass reduces the effective amount of biomass available for burning per unit mass of the sample, and thus the net amount of energy extracted.

Table 2.1 shows how the energy in solid biomass compares with other solid, liquid and gaseous fuels.

Table 2.1. Comparison of energy content of solid biomass with other solid, liquid and gaseous fuels (data source: [11]).

Fuel	Specific energy content (MJ kg^{-1})	Energy density (MJ L^{-1})	Comments
Woody biomass	8–20	13	Energy content (heating value) depends on moisture content and density
Coal	26–35	34–49	Four types, depending on carbon content: Anthracite (86–97% carbon, highest heating value); Bituminous coal (45–86% carbon); Sub-bituminous coal (35–45% carbon); Lignite (25–35% carbon, lowest heating value)
Ethanol	30	21	
Coconut oil (CNO)	39		
Petrol (gasoline)	46	34	
Liquid petroleum gas (LPG)	46	26	A blend of butane and propane, liquified under modest pressure in domestic cylinders
Diesel	48	36	
Methane	55	0.037	At standard temperature and pressure (STP)

2.3 Forms of solid biomass feedstock

In what form is biomass available as an energy feedstock?

Biomass for use as an energy feedstock can originate from a multitude of sources. The primary biomass source is woody and herbaceous plant materials. Other related classes of materials are forestry and agricultural crop residues. But the list extends to all other organic residues of agricultural or industrial processing, and also includes food processing residues, animal wastes and urban wastes including yard wastes, municipal solid wastes and sewage wastes. Indeed, any material that contains an organic component forms a likely candidate for the production of energy. This includes waste products consisting of a mixture of inorganic and organic substances.

The most obvious biomass sources are dedicated energy crops which are planted for the explicit purpose of producing energy, as well as by-products of industrial and manufacturing processing of woody biomass. Many sources, however, have a mixed composition, where not all of the material is organic. This heterogeneous collection of sources may be conveniently divided into plant sources and mixed (animal, plant and inorganic) sources. Table 2.2 lists some of the main categories of biomass energy sources with relevant examples.

Table 2.2. Sources of solid biomass energy feedstock (data source: [12]).

Biomass category	Examples
Plant sources:	
Dedicated energy crops	Herbaceous energy crops Short-rotation woody crops
Algae	Microalgae, macroalgae (seaweed), blue-green algae
Forestry residues	Timber logging residues
Industrial residues	Sawdust, hog fuel, bagasse
Agricultural residues	Corn stover, wheat straw, rice straw, sugarcane trash, coconut shells, coconut husks
Mixed sources:	
Food industry residues	Expired supermarket food
Animal wastes	Animal manure, animal fat
Urban wastes	Yard wastes (prunings and trimmings of gardens and hedges, lawn mowing), industry wastes, municipal market wastes
Domestic wastes	Municipal solid waste
Sewage	Solid/liquid sewage

2.3.1 Dedicated energy crops

Dedicated energy crops can come in a wide variety. These are typically divided into herbaceous (i.e. grassy) and short-rotation wood crops. The former are perennial grasses (i.e. grasses that can live for more than a few years) that are harvested annually after they have reached their age of maturity, while the woody crops are hardwoods that can be harvested after five to eight years of planting.

Examples of the herbaceous species in Europe and the USA are miscanthus, switchgrass, reed, canary grass, Cynara, bamboo, sweet sorghum, kochia and wheatgrass. Short-rotation woody crops include poplar, willow, eucalyptus, locust, maple, cottonwood, green ash black walnut and sycamore.

2.3.2 Forestry and industrial residues

While some sources of biomass energy may be grown as dedicated energy crops, much of this energy feedstock can be derived from the timber and pulp and paper industries. **Forestry residues** are the portion of trees discarded in the forest after logging has been completed. These residues include the tops, branches and leaves of logged trees as well as (the unharvested) dead or diseased trees.

The processing of logs during timber milling provides another source of energy feedstock. **Industrial residues** from timber milling include bark, sawdust, off-cuts, shavings (collectively known as hog fuel) produced from the industrial processing of logs.

Black liquor from the **pulp and paper industry** provides yet another source of energy derived from biomass. This is formed during the extraction of cellulose from wood in the paper manufacturing industry. It is the liquid that remains after the removal of cellulose fibres for paper formation, and consists of lignin and other chemicals used in the extraction process. It provides a significant source of fuel for the pulp industry, and is used for combustion and gasification [13].

An important industrial residue that results from the processing of an agricultural product is **bagasse,** which is produced as a residual product in the **sugarcane industry**. This is the ligno-cellulosic residue that remains after crushing of the sugarcane stalk for extraction of sugarcane juice, and is used by most sugar mills world-wide as a means of generating process heat and electricity to power the milling operations.

2.3.3 Municipal solid waste

Household and other trash generated in towns and cities (collectively termed municipal solid wastes (MSW)) provide an important source of organic matter that can be used for various energy needs. One example is the use of the organic fraction of MSW in power generation via biogas production, pyrolysis and incineration in waste-to-energy (WTE) plants. In 2018, there were 492 WTE plants in Europe, 96 of which were using thermally-treated waste [14].

2.3.4 Animal waste

Solid animal wastes such as cow dung or pig manure are readily-available energy sources for cooking and heating in developing countries. This source may be either used directly for fuelling fireplaces, enhanced through the addition of herbaceous biomass before use, or used as the substrate for the production of biogas for domestic and commercial applications.

2.3.5 Sewage

Domestic sewage discharge comprises a high percentage of organic matter, and provides an energy feedstock for the production of other forms of energy. An example is the use of biochemical techniques such as digestion for the production of biogas for power generation and other uses. The anaerobic conditions inside municipal sewage systems provide ideal conditions for the generation of methane. As this gas is a potent greenhouse gas, special care has to be taken at municipal sewage systems to ensure that such naturally-produced products do not escape into the atmosphere. Flaring (burning the methane before releasing the products to the atmosphere) is commonly used to eliminate this danger.

(See question 2.1—Sources of biomass in your region.)

2.4 Energy uses of solid biomass

We have seen that a wide variety of biomass sources is available as energy feedstock in any country. It is natural to ask how such resources are actually used as energy sources.

The energy uses of solid biomass range from cooking and heating at the domestic level to provision of process heat for industrial processes and the generation of electrical power for large grid systems at the national level. As we will see in later chapters, solid biomass also provides the initial feedstock for the production of secondary fuels which may be either solid, liquid or gaseous in nature. Examples include the pyrolysis and gasification processes, and the industrial production of biogas from solid feedstock such as expired supermarket food and other organic industrial wastes.

Solid biomass provides the primary source of **cooking fuel** for most of the world's poor. This is because (in its original, or **traditional** form) it is cheap and/or easily available in rural areas of countries, (where the poor usually reside). We will see in greater detail in chapter 12 how the use of traditional biomass as cooking fuel and poverty are intricately linked. We noted in section 2.2.3 that biomass has some distinctive disadvantages over other fuel sources because of its moisture content and physical form. In section 2.7, we will see how the fuel properties of woody and herbaceous biomass may be improved through pre-treatment and densification procedures.

The **industrial processing** of consumer products often requires **process heat** for product manufacture. Examples include the drying of copra in the coconut oil industry, processing sugar juice in the sugarcane industry and the manufacture of food items in food manufacture. Residues from the industrial processing of the relevant crops often provides a readily-available source of energy for this purpose. In the **power generation sector**, solid biomass can readily replace coal in **thermal power plants**.

2.4.1 Case study: biomass energy use in the USA

The use of biomass as an energy feedstock in the USA provides a quantitative case study of the importance of this energy resource in modern economies. In the year 2020, biomass provided 4532 trillion British Thermal Units (i.e. 4780 PJ) of energy towards the primary energy consumption requirements of the United States of America [15]. This amounted to 4.9% of the Total Primary Energy Consumption of the nation. The pie chart in figure 2.8 shows how this source of energy was shared amongst the country's biomass energy sub-sectors.

In the USA, biomass is used by several economic sectors, ranging from the residential to the industrial. Table 2.3 shows how this energy use is distributed amongst these sectors.

In the table:

- the **industrial sector** consists of the manufacturing, agriculture, forestry, fishing and hunting and hunting sectors;
- the **transportation sector** consists of all transport vehicles;

- the **residential sector** comprises all private households which use energy for space heating, water heating and cooking;
- the **power generation sector** consists of condensing and combined heat and power (CHP) plants which use biomass to fire their furnaces; and
- the **commercial sector** comprises businesses, the Federal State, local government, other private and public organisations.

(See exercise 2.2—Use of biomass energy in the economic sectors of your country.)

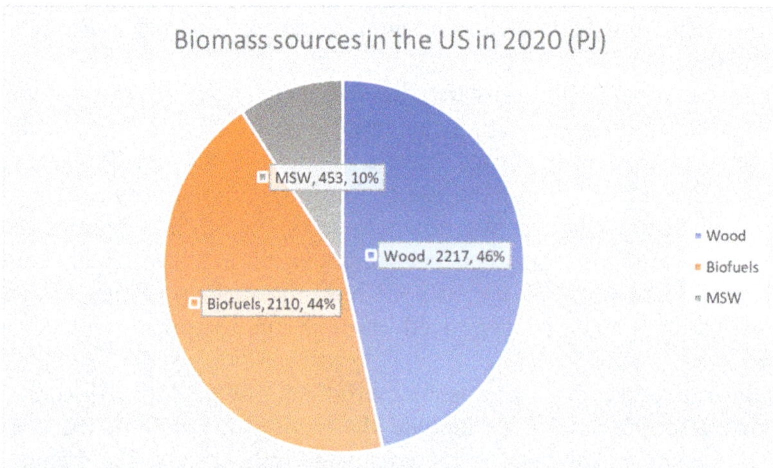

Figure 2.8. Sources of biomass energy in the USA in 2020 (data source: [15]).

Table 2.3. Biomass energy use sectors in the USA—2020 (data source: [15]).

Sector	Energy use (BTU)	Energy use (PJ)	Share of total biomass energy (%)
Industrial	2246×10^{12}	2370	50
Transportation	1263×10^{12}	1332	28
Residential	458×10^{12}	483.2	10
Power generation	424×10^{12}	447.3	9
Commercial	141×10^{12}	148.8	3
Total	$\mathbf{4532 \times 10^{12}}$	**4781.3**	**100**

2.5 The important properties of woody biomass

Our main interest in biomass is as an energy feedstock (i.e. as a source of energy). Thus, an essential property of woody biomass is its energy per unit mass or volume. This is variously called its **energy content**, **heating value** or **calorific value**.

As indicated in section 2.2.3, the presence of moisture in biomass affects the value of its energy content as well as the efficiency with which it is used. The ability to ignite biomass, and the quality of the flame produced both depend on the presence of moisture within the structural matrix of the biomass feedstock. **Moisture content** is therefore an important property of biomass for consideration.

The solid matter contained in a sample of biomass has a lower density than that of water. We thus see that the presence of water in biomass is a determining factor in the overall density of the biomass. **Density** is therefore another property that must be taken into consideration when assessing the quality of biomass as an energy source. It has a direct bearing on the transportability, storage and general handling of biomass.

Moisture content, density and energy content are thus three important parameters that determine the efficacy of woody and other forms of biomass as energy feedstocks. To proceed further, one needs to provide definitions for these terms that are practical and scientifically sound. Prior to 2001, a wide variety of terminologies existed for the classification of fuelwoods and other forms of biofuels. To bring some degree of uniformity amongst these classifications, the Food and Agriculture Organization (FAO) of the United Nations met with other institutions in Rome in October 2001 to produce what became known as the **Unified Bioenergy Terminology (UBET)** [16]. The definitions for moisture and energy content given below are the same in form as those proposed in the FAO paper, although the symbols have been altered for clarity and ease of use.

2.5.1 Moisture content

The **moisture content** of biomass is measured by firstly determining the initial mass of a sample of biomass, oven drying it for 24 h and determining the resulting (dry) mass. One of the two alternative definitions shown below is then used to determine the appropriate type of moisture content.

The **moisture content (dry basis)** is given by

$$W_{\mathrm{D}} = \frac{100(m - m_{\mathrm{D}})}{m_{\mathrm{D}}}\% \tag{2.3}$$

where W_{D} is the percent moisture content (dry basis), m is the mass of the sample (before drying), and m_{D} is the mass of the sample after oven drying at 105 °C for 24 h.

The **moisture content (wet basis)** is given by

$$W_{\mathrm{W}} = \frac{100(m - m_{\mathrm{D}})}{m}\% \tag{2.4}$$

where W_{W} is the percent moisture content (wet basis) and m, m_{D} have the same definitions as in expression (2.3).

(See exercise 2.3—Calculating the moisture content of wood.)

2.5.2 Energy content

Before providing a definition for the energy content of biomass and biofuels, it is important to note that this quantity is defined using a range of terminologies. Stated briefly, the **energy content**, **heating value** or **calorific value** refer to the same energy-related parameter of biomass, and are further classified into **upper** or **lower heating values** in the case of the heating value, and into **gross** or **net calorific values** in the case of the latter expression. For all intents and purposes, one can equate the upper heating value to the gross calorific value, and the lower heating value to the net calorific value.

Confining our attention to calorific values, the difference between the gross and net values depends on the amount of heat of combustion that is actually available for use. This is most readily seen by considering how the calorific value is measured.

To determine the calorific value of a substance, a small sample is placed in a device known as a **bomb calorimeter**, and burnt completely in an oxygen-rich atmosphere. Assuming that the sample consisted of fuel containing carbon, oxygen and hydrogen only (e.g. the substance is a carbohydrate), the products formed will be water and carbon dioxide only. Now the energy user is interested in the amount of energy available for use. (In the case of a vehicle owner buying fuel for a vehicle, this is the amount of energy available to drive the vehicle.)

However, the water formed as a product of combustion will be in its gaseous form (steam). As the water is in the evaporated form, it is reasonable to say that some of the heat of combustion was used up as the heat of evaporation of the water. Therefore, to determine the full amount of the heat of combustion produced in the oxidation reaction, the amount of heat required to evaporate the water from its state at ambient temperature (i.e. its latent heat of vaporisation L_v and the heat ΔE required to raise its temperature to the boiling point) would need to be taken into account.

One notes, however, that this additional amount of energy will not be available for useful purposes (it escapes with the exhaust in the case of a vehicle), and therefore the useful calorific value will always be less than the total heat of combustion. The Net Calorific Value (NCV) is calculated by subtracting the additional heat necessary for vaporisation from the total heat of combustion. While the Gross Calorific Value (GCV) is the larger amount, which includes the energy used to evaporate the water, the NCB refers to the actual amount of energy available for use.

The UBET-derived formula for the **Net Calorific Value** is

$$E(W) = \frac{E_0(100 - W)}{100} - \frac{2.44W}{100} \tag{2.5}$$

where W is the moisture content (%), $E(W)$ is the Net Calorific Value (or Lower Heating Value) of the biomass or biofuel (MJ kg^{-1}), E_0 is the NCV of the bone dry sample (i.e. sample with $W = 0\%$). The number 2.44 is the energy (MJ kg^{-1}) required to totally evaporate 1 kg of water, initially present at the ambient temperature.

Note that the above formula may also be read as

Net Calorific Value = Gross Calorific Value – Heat of Vaporisation of Water

The NCV will also depend on the hydrogen content of the biomass. This (relatively small effect) is not accounted for in the above formula.

(See exercise 2.4—The effect of moisture content on the net calorific value of wood.)

(See exercise 2.5—A 'theoretical' energy–moisture curve.)

The density of woody biomass needs further elaboration and is considered fully in section 2.6.

2.6 Energy–moisture relations

According to expression (2.5), the energy content of biomass depends on its moisture content, its value reducing with increase in the latter quantity. This relationship between moisture and energy content seems to be a reasonable one if one adopts a simple model of woody biomass where the total mass consists of $W\%$ water with the solid biomass contributing the rest of the mass. According to such a model, for a sample with a fixed total mass, as the moisture content increases, the mass due to the biomass in the sample will reduce, leading to a reduction in the total energy content of the sample.

This theoretical trend is indeed exhibited by the experimental energy–moisture relationships of all woody plant species, which reveal a general decline in their energy contents as the moisture content increases.

2.7 Density–moisture relations

2.7.1 The density of wood

The density of a solid may be simply defined as its mass per unit volume. This, however, does not specify the method of determining the density, and issues arise when the volume of the sample being used in the determination changes due to environmental factors such as temperature and relative humidity that affect the properties of the sample. Wood is subject to such variations, and therefore the measurement of its density needs special attention.

The density of wood is usually determined in terms of its specific gravity and moisture content (dry basis). In the forestry literature (see [17]), the density is usually tabulated numerically in terms of the specific gravity and the moisture content.

The moisture content of wood depends in an intricate way on the ambient temperature and relative humidity ([17], chapter 4). This moisture is present as **free water** in the cavities inside cells (called cell lumen) and as **bound water** in the cell walls. As freshly-cut green wood dries, the free water is lost first, leaving the volume of the sample unchanged. The point at which all free water is lost is called the **Fibre Saturation Point (FSP)**.

Any further moisture loss occurs at the expense of the bound water from the cell walls. The wood sample now begins to lose volume, a phenomenon known as **shrinkage**. The behaviour of the density change with moisture reduction can now be quite different from its earlier behaviour. These changes in moisture content have significant impact on the mechanical properties of wood. The reader is referred to the Wood Handbook [17] and other authoritative sources for further readings on this subject.

To take into account these moisture content-related effects, the density of wood is usually defined in terms of a special parameter called the **basic specific gravity G_b**.

This is given by

$$G_b = \left(\frac{m_D}{V_G}\right)/\rho_w \tag{2.6}$$

where m_D is the oven-dry mass of the sample, V_G is the green volume of the sample and ρ_W is the density of water.

2.7.2 Density–moisture curves

Simpson [18] provides equations for the density of wood that take into account its shrinkage as follows. He first describes the loss in volume after the FSP in terms of the **shrinkage (S)** given by

$$S = aS_t \tag{2.7}$$

where $S_t = 0.265G_b$ is the total change in volume between the FSP and zero moisture content and $a = \frac{30 - W}{30}$, with FSP assumed to occur at $W = 30\%$ (dry basis).

He then uses these definitions to formulate equations describing the density–moisture relations of wood above and below the fibre saturation point as follows.

Above FSP, the density–moisture relation is given by

$$D = \rho_w G_b\left(1 + \frac{W}{100}\right) \tag{2.8}$$

where D is the density of the wood, ρ_w is the density of water, G_b is the basic specific gravity and W is the moisture content.

Below FSP, the density is given by

$$D = \frac{\rho_W G_b\left(1 + \frac{W}{100}\right)}{1 - 0.265a\, G_b}. \tag{2.9}$$

(See exercise 2.6—Density–moisture curves.)

2.8 Pre-treatment and densification of biomass

How can we improve the fuel properties of biomass?

We have noted above that the fuel properties of biomass are adversely affected by its moisture content. Thus, reduction of this parameter is an important requirement for the improvement of the fuel properties of solid biomass. Several forms of biomass in their natural state have low bulk densities, which affect their handling and storage. There is, therefore, a need to densify the fuel before use. This is assisted by certain pre-treatment procedures. These processes are described in more detail below.

2.8.1 The need for densification

We saw in section 2.2.3 that compared to other fuels, biomass has the disadvantage that it has relatively low energy content and high moisture content. The available forms of biomass include grasses and agricultural residues. These have the added disadvantage that they have low bulk densities.

A bundle of (say) wheat straw consists of individual straws separated by air gaps, which increases the overall volume (i.e. the bulk volume) of the bundle. This leads to an effective density (the bulk density) that is lower than the density in the absence of the air gaps. Typical examples are shown in table 2.4.

Table 2.4. Typical bulk densities of some agricultural residues compared to that of woody biomass and coal (after [19]).

Agricultural residue	Typical bulk density (kg m^{-3})
Wheat straw	20
Corn stover	60
Soybean hull	170
Corn cobs	270
Hardwood	320
Coal (lignite)	700

The low bulk density of agricultural residues leads to low energy density, making it more expensive to transport from the site of production to its final destination. More generally, biomass feedstocks also suffer from their irregular shape and size, all of which hinder the handling, transportation and storage of this fuel, and adds to the costs of energy production.

These disadvantages are partially addressed through **densification** of the fuel, either through mechanical or thermal means. This process helps by:

- increasing the energy density of the fuel;
- improving the combustion properties of the biomass;
- simplifying the mechanical handling, transportation and storage facilities; and
- reducing overall costs.

2.8.2 Pre-treatment of biomass

Before densification can proceed, however, the biomass fuel requires pre-treatment. This preliminary step assists in facilitating densification through:

1. Chop-length selection and grinding:

   The size of the biomass (called the particle size) needs to be cut to the size appropriate for the densification method and/or its grinding into finer particles for subsequent treatment.

2. Drying:

   Lowering the moisture content increases the energy content of the fuel as well as preparing it for densification. The ideal moisture content should lie in the 8–20% range.
3. Addition of binding agent:

   One of the densification techniques is pellet formation from finely ground biomass. These pellets are often difficult to form due to their tendency to fall apart. A binding agent needs to be mixed with the initial ground material to hold it together.
4. Steaming:

   Subjecting the biomass to steam treatment releases natural binding agents such as oils and starch present in the biomass.
5. Torrefaction:

   This is a mild form of pyrolysis (heat treatment in the absence of oxygen) that is carried out at temperatures lower than the usual pyrolysis processes, and produces a hydrophobic solid that has lower moisture content than the initial biomass [20].

2.8.3 Densification techniques

The main methods of densification may be divided into the two distinct categories of **mechanical** and **thermal** densification. In the former, mechanical pressure is applied to the biomass to increase its density [20]. Thermal densification consists of various forms of heat treatment to improve the fuel properties of the biomass feedstock, and includes slow and fast pyrolysis as well as torrefaction.

2.8.3.1 Mechanical densification

Methods for mechanical densification include processing the biomass into **bales**, **pellets**, **cubes**, **briquettes** and **pucks** [19]:

1. Bale production:

   This is the traditional method used for the on-farm processing of agricultural residues after crop harvest, and involves the use of a baler [19].
2. Pellets:

   These are formed through the process of extrusion after the biomass has been ground to a suitable particle size. Various extrusion technologies are available [20].
3. Cubes:

   These are larger versions of pellets that are formed through the processes of compression and extrusion [19].
4. Briquettes:

   These are large pellets formed through the process of punching or screw extrusion.
5. Pucks:

   These are produced from material that has first been subjected to treatment using a briquetter. The output is then sliced into portions that resemble hockey pucks.

6. Wood chips:
   An additional form in which solid biomass is used as energy feedstock.

The reader is referred to [19] for pictorial illustrations of all products of mechanical densification mentioned above.

2.8.3.2 Thermal densification

In contrast to mechanical techniques of densification, subjecting biomass feedstock to thermal treatment under controlled conditions brings about chemical change that adds to the overall fuel properties. At the lowest end of the temperature scale, the pre-treatment technique of torrefaction may be sufficient in itself to provide ample improvements.

Torrefaction is carried at temperatures of 280–320 °C in oxygen-deficient atmospheres for short periods (a few minutes) to yield fuel that has lower moisture content and is hydrophobic. The former property increases the energy density of the fuel while the latter improves its storage properties by making it less susceptible to biological degradation. The cost of the heat treatment is fortuitously reduced, as up to 80% of the heating fuel is provided by the volatile compounds produced through the heat treatment.

Slow pyrolysis is carried out at the higher temperature range of 350–500 °C and for periods of a few hours in an oxygen-free environment [19]. The resulting fuel is mainly in a solid form (charcoal).

The technique of **fast pyrolysis** is a much more rapid technique which is carried out for only a few seconds within the temperature range of 400–500 °C. The result is the production of a liquid product, called bio-oil, in addition to the solid charcoal. The production of bio-oil provides the additional storage and handling advantages through the ability to transport and store this liquid fuel in tanks. These techniques of pyrolysis are treated in detail in the next chapter, which investigates the thermochemical changes that occur in biomass under heat treatment.

(See question 2.2—The advantages of densification.)

Exercises, questions and projects

Exercises

Exercise 2.1. Anatomy of woody plants.

Describe the anatomy of the stem of a woody plant, clearly identifying the various plant tissue systems present and their functions. State the particular roles these tissues play in the transport of water and minerals from the soil, as well as the sugars produced in the leaves of the plant.

Exercise 2.2. Use of biomass energy in the economic sectors of your country.

Table 2.3 shows how biomass energy is utilised by the various economic sectors in the USA. By using data available from official sources or otherwise, estimate the percentage share of the total available biomass used by the same group of economic sectors in your country (or state if you are living in the USA). How is the biomass energy distributed within the sub-sectors of your industrial sub-sector?

Exercise 2.3. Calculating the moisture content of wood.

A block of wood 45.0 cm^3 in volume taken from a freshly-cut branch of a woody plant weighed 0.050 kg. The same mass reduced to 0.032 kg after it was oven-dried for 24 h at a temperature pf 105 °C. Determine the dry and wet basis moisture content of this wood.

Exercise 2.4. The effect of moisture content on the net calorific value of wood.

If the net calorific value of the bone dry sample of the wood in exercise 2.3 was 20 MJ kg^{-1}, use equation (2.5) to evaluate its net calorific value before and after drying.

Exercise 2.5. A 'theoretical' energy–moisture curve.

Use equation (2.5) to calculate the net calorific value of the wood sample for values of the moisture content between 10% and 50% at intervals of 10%. Use these values to plot a 'theoretical' energy–moisture curve for the wood by plotting a graph of the calorific value (on the y-axis) against the moisture content (on the x-axis).

Exercise 2.6. Density–moisture curves.

Tabulate the values of the densities of the woody plant material considered in exercise 2.3 at moisture contents of 5%, 10%, 15%, 20%, 30%, 40%, 50% and 60% (wet basis), and G_b of 0.35 and 0.65. Plot the curves of density against the moisture content for each value of G_b and comment on the nature of the curves.

Questions

Question 2.1. Sources of biomass in your region.

Identify all the sources of solid biomass available in your town, municipality or region, and group them according to the categories described in table 2.2. Comment on the quality of each of these energy sources.

Question 2.2. The advantages of densification.

In no more than 200 words, explain how densification improves the fuel properties of solid biomass, and how pre-treatment prepares biomass for more effective densification.

(i) Explain, with the aid of examples, how densification leads to an increase in the energy content of solid biomass.

(ii) A company obtains woody biomass and crop residues from a large farm that produces wheat and lumber, and transports it several kilometres to a bulk-store for sale and distribution. Use a hypothetical case study to show quantitatively how densification of the transported biomass can lead to cost savings for the company.

Projects

Preparing a research proposal

Solid biomass, especially the residual biomass of industry and waste products of municipal/community activities, may provide a lucrative source of energy.

You are interested in carrying out a detailed investigation of the nature and amount of solid biomass that is being produced by your community or country, and wish to apply to the relevant authorities for possible funding for your research project.

To achieve your aim, you will first need to prepare a project proposal for submission to the authority. This proposal will

- state what you want to do;
- show how you will carry out the investigation; and
- provide an estimate of how much it will cost and other information such as how long it will take and what other requirements it will need to satisfy.

But to improve the chances of success, you will also need to

- provide a motivation/justification for the project, and
- find out what information already exists, and whether someone else has already carried out similar work.

The elements of your research proposal should therefore consist of the following:

1. An **aim/objective** that states the general aim of the project and specifies each specific objective of the task.
2. An **introduction** that shows the motivation and justification of the project (i.e. why you want to do it).
3. A **literature/internet review** of the topic that investigates what work has already been done by others.
4. A **methodology** which describes how you will carry out the project.
5. A short summary of the **expected outcomes** of the work.
6. A **detailed budget** of the costs involved.

What you must do:

To complete this task, you will need to firstly do a literature/internet search on what is already known or done towards the aims/objectives of this project. Once you have completed this preliminary requirement, you will be in a position to complete the main segment of this task, which is to **produce a short report (of no more than 1000 words) according to the headings identified in the list above**. At this stage, the budget does not need to be a very detailed one.

You will begin by choosing a suitable title for your research proposal!

References

[1] What are the products of photosynthesis. ThoughtCo. [accessed 2 November 2022]. Available from https://thoughtco.com/the-products-of-photosynthesis-603891

[2] Basics of reaction profiles. Chemistry LibreTexts [accessed 2 November 2022]. Available from https://chem.libretexts.org/Bookshelves/Physical_and_Theoretical_Chemistry_Textbook_Maps/Supplemental_Modules_(Physical_and_Theoretical_Chemistry)/Kinetics/06%3A_Modeling_Reaction_Kinetics/6.03%3A_Reaction_Profiles/6.3.02%3A_Basics_of_Reaction_Profiles

[3] Tissues, organs and organ systems. Khan Academy [accessed 2 November 2022]. Available from https://khanacademy.org/science/biology/principles-of-physiology/body-structure-and-homeostasis/a/tissues-organs-organ-systems#:~:text=Major%20organ%20systems%20of%20the%20human%20body%20,thyroid%2C%20parathyroids%20...%20%208%20more%20rows%20
[4] Wikipedia [Plant cell structure]. Wikipedia [accessed 2 November 2022]. Available from https://commons.wikimedia.org/wiki/File:Plant_cell_structure.png
[5] Structure of lignocellulose. Scitable [accessed 2 November 2022]. Available from https://nature.com/scitable/content/structure-of-lignocellulose-14464273/
[6] Hemicellulose. Wikipedia. 2007 [accessed 2 November 2022]. Available from https://en.wikipedia.org/wiki/Hemicellulose
[7] Biochemistry. Wikipedia. 2009 [accessed 2 November2022]. Available from https://en.wikipedia.org/wiki/Biochemistry
[8] Cellulose. Wilipedia. 2007 [accessed 2 November 2022]. Available from https://en.wikipedia.org/wiki/Cellulose
[9] Xylose. Wikipedia. 2009 [accessed 2 November 2022]. Available from https://en.wikipedia.org/wiki/Xylose
[10] Lignin. Wikipedia. 2009 [accessed 3 November 2022]. Available from https://en.wikipedia.org/wiki/Lignin
[11] Basics: Data and Statistics. EIA [accessed 3 November 2022]. Available from : https://eia.gov/energyexplained/coal/
[12] Biomass resources. EERE [accessed 3 November 2022]. Available from https://energy.gov/eere/bioenergy/biomass-resources
[13] Wood energy. USDA [accessed 3 November 2022]. Available from https://wood-energy.extension.org/wood-processing-residues/
[14] Waste-to-Energy Climate Roadmap. CEWEP [accessed 3 November 2022]. Available from www.cewep.eu
[15] Biomass-renewable energy from plants and animals. US Department of Energy EIA [accessed 3 November 2022]. Available from https://eia.gov/energyexplained/biomass/
[16] Unified Bioenergy Terminology(UBET). Food and Agriculture Organization of the UN (FAO). 2004 [accessed 3 November 2022]. Available from http://fao.org/3/j4504E/j4504e00.htm
[17] Forest Products Laboratory, USDA 2010 *Wood Handbook—Wood As An Engineering Material* (Madison, WI: USDA) p 508
[18] Simpson W T 1993 Specific gravity, moisture content and density relationships for wood *General Technical Report FPL* (GTR-76. USDA, Forest Service, Forest Products Laboratory) 13
[19] Biomass densification for energy production. Ministry of Agriculture, Food and Rural Affairs, Ontario. June 2011 [accessed 3 November 2022]. Available from https://ontario.ca/page/biomass-densification-energy-production
[20] Tumuluru J S 2020 *Biomass Densification – Systems, Particle Binding, Process Conditions, Quality Attributes, Conversion Performance and International Standards* (Cham: Springer) p 191

IOP Publishing

Anirudh Singh

Chapter 3

Thermal and thermochemical conversion of solid biomass

Biomass is generally not available in a form that is convenient for use as an energy source. With the notable exception of combustion, it generally needs to be converted to a secondary fuel before use. This chapter begins with an overview of all such conversion processes, which can be chemical, thermochemical, biochemical or electrochemical in nature. It then provides a detailed treatment of the thermochemical process, which consists of processes of combustion, pyrolysis and gasification.

The nature of combustion is first described, and its chemical stoichiometry in air is elaborated on. The several processes (which occur in temperature zones) involved in the combustion of solid biomass are then highlighted, and the solid, liquid and gaseous by-products noted. The process of pyrolysis (which occurs in the absence of oxygen) is next described in detail and the various forms, ranging from very slow to flash pyrolysis, are noted. The chapter ends with a study of the gasification of biomass, which occurs in thermal treatment of biomass under reduced oxygen supply and the controlled presence of steam to produce syngas.

3.1 Introduction

The energy in biomass can be put to various economic uses. As we saw in chapter 2, in the USA these typically involve the industrial, commercial, residential, transport and power generation sectors.

Biomass is not always suitable as an energy source in its original form, and generally requires a conversion process. In the case of cooking, heating or power generation, this fuel may be used with only minor modifications (usually physical in nature). However, in applications such as transportation, it needs to be converted to

doi:10.1088/978-0-7503-3555-3ch3

either liquid or gaseous forms before use. In addition, it may need to meet further requirements that are specific to engine types. For example, diesel and gasoline engines require quite different types of fuels for their operation. For biomass to be of use for either of these purposes, it must be converted into a liquid biofuel that satisfies the specific requirements of a fuel.

It is clear, therefore, that in general, biomass needs to undergo conversion before it can be put to effective use. There is a variety of ways in which this can be accomplished. The aim of this and the following chapter is to provide a first introduction to some of the most important of these pathways.

This chapter begins with a diagrammatical illustration of these possible conversion processes. It then investigates the three inter-related **thermochemical processes** of **combustion**, **pyrolysis** and **gasification** in more detail.

Combustion of solids in air can be either complete or incomplete. The conditions for complete combustion are reviewed by considering the associated stoichiometry of the oxidation reaction. In the specific case of solid biomass, combustion occurs in temperature zones within which distinctly different chemical processes occur. Specific attention is given to the by-products of combustion, including aerial pollutants that pose serious environmental hazards.

The pyrolysis of solid biomass is introduced by first describing the pyrolysis process, and noting the physical parameters within the pyrolysis reactor that maximise product yields. The four types of pyrolysis (from very slow (taking days or weeks) to fast and flash pyrolysis (which takes seconds)) are described and the relative yields of the products noted. This is followed by an overview of the various types of industrial and experimental pyrolysis reactors in use today.

The subject of gasification of solid biomass is introduced with a description of the gasification process. In common with the other two thermochemical processes, this is also characterised by temperature zones inside the gasifiers (the reactors used for gasification). The chemistry of gasification is elucidated and the list of types of commercial gasifiers available today is reviewed. The chapter ends with a summary of the three thermochemical processes and their products.

3.2 Pathways for biomass conversion

3.2.1 Overview of biomass conversion processes

In cases such as cooking, space heating or power generation where the requirement is the generation of heat, the conversion process is the combustion of the biomass itself. In other cases, the biomass may need to be converted into **secondary fuels** such as those needed for diesel or biogas-driven vehicles and power plants. The possible processes by which all such transformations are carried out may be categorised into **thermal**, **thermochemical**, **chemical**, **biochemical** and **electrochemical**. Figure 3.1 provides a pictorial representation of the pathways by which these processes can proceed.

This chapter provides an introduction to the science and industrial significance of the thermal and thermochemical processes of **combustion**, **pyrolysis** and **gasification**, beginning with an overview of all three.

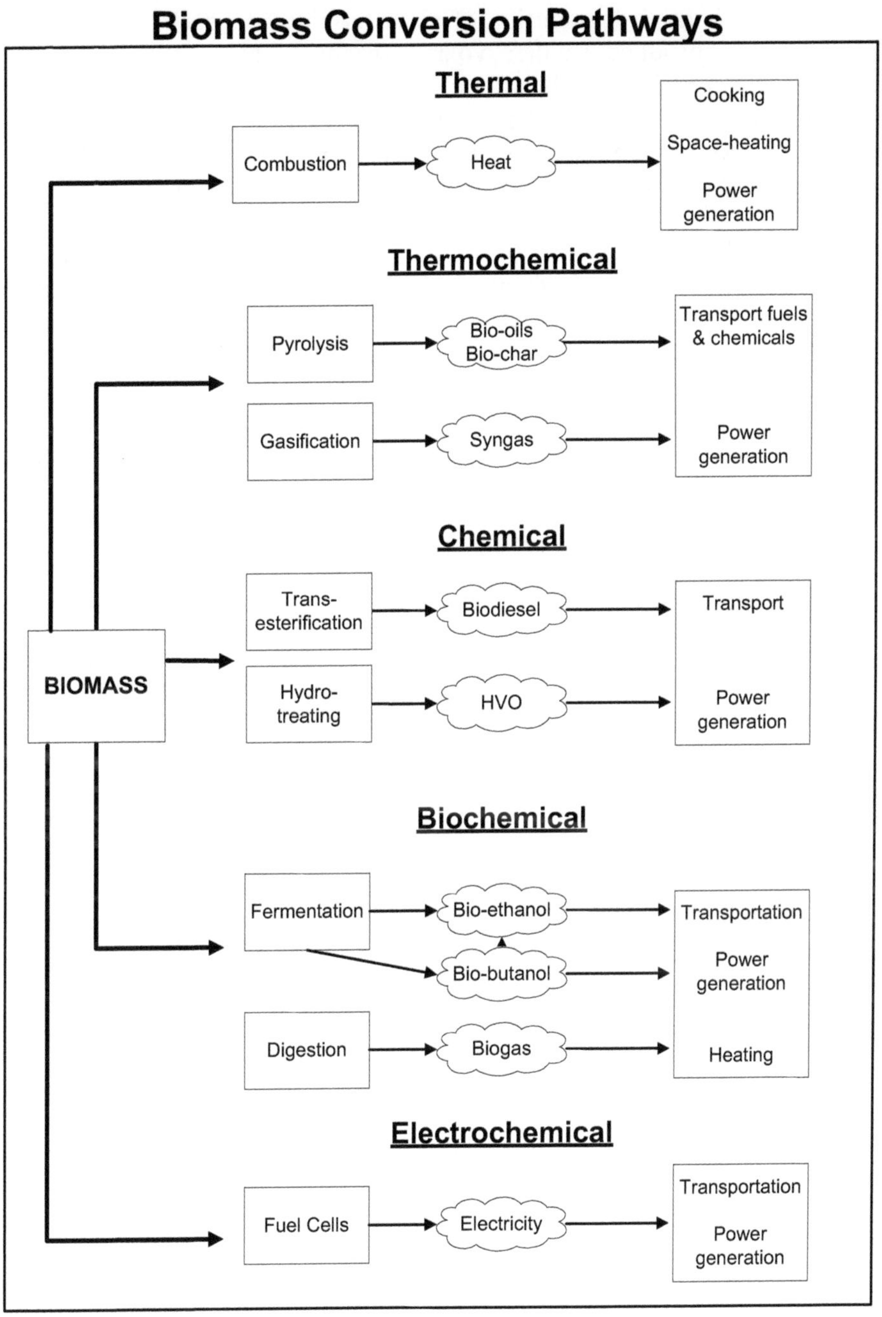

Figure 3.1. Pathways for conversion of biomass for end use.

3.2.2 Thermochemical processes

Three related thermochemical processes of combustion, pyrolysis and gasification occur when solid organic matter is subjected to thermal treatment. These processes range from the thermal degradation of giant biopolymers such as cellulose and hemicellulose to produce smaller volatile products and solid carbonaceous products, to the complete combustion of the organic matter into carbon dioxide, water and other gaseous oxides. The final products range from a mixture of liquid bio-oils, solid bio-char and gaseous hydrogen and carbon monoxide on the one hand, to a predominantly gaseous mixture of the last two energy-rich compounds on the other.

The chemical processes that occur are complex and depend on both the starting material (called the **feedstock** or **fuel**) as well as the process parameters. These processes are still the subject of active research and investigation in many university, industry and government-owned laboratories. The products formed are of considerable commercial value and provide the basis for several industrial processes.

The following section discusses the combustion process in detail, beginning with its nature and chemistry.

3.3 Combustion

3.3.1 Nature of combustion

According to the earliest theories of combustion [1], matter was comprised of a substance called **phlogiston**. Combustion was the process in which phlogiston was released (leaving behind the ashes). This theory could not withstand the scientific investigations of the early scientists such as Antoine Lavoisier of France and Joseph Priestly of England, and by 1798, Benjamin Thompson was able to link the phenomenon of combustion with the modern scientific concepts of heat, thermodynamics and the kinetic theory of gases [2].

Combustion, or **burning**, is the chemical reaction between a substance, called the **fuel**, and its **oxidant** (usually oxygen). We saw in chapter 2 that chemical reactions can be either endothermic (requiring energy input to proceed) or exothermic (giving out net energy to the environment, usually in the form of heat). Combustion is an exothermic process. Apart from the material products, the output of the process includes a net release of energy, mostly in the form of heat.

Combustion is not always visible. When an organic solid is subjected to an external heat source, it produces volatile flammable products through an endothermic process. Both the solid and the volatile products emit radiation in the form of blackbody radiation. The wavelength spectrum of this radiation depends on the temperature, and when it is high enough, a portion of it appears in the visible region. This results in the visible glowing of the solid and the flaming of the gaseous components, respectively. This release of light is called incandescence.

3.3.1.1 Complete and incomplete combustion

Combustion is **complete** when all the fuel has been oxidised completely to its highest oxide. In the case where the fuel is a hydrocarbon (where the molecules contain carbon and hydrogen atoms only), this implies that the only material products will be carbon dioxide and water. When there is sufficient or excess oxygen available, the result will therefore be carbon dioxide, water and energy. In the example of methane gas, the process is described by the equation:

$$CH_4 + O_2 = CO_2 + H_2O + \Delta H_c \quad (3.1)$$

where ΔH_c is the enthalpy change or heat of combustion, (i.e. the heat given out in this exothermic reaction).

The amount of oxygen present in the above oxidation reaction is said to be **stoichiometric** if it is exactly equal to the amount required for the reaction to continue to completion without any residual reactants remaining in the final product. If oxygen is not available in sufficient quantity, then the products will in general contain other substances as well, including gaseous carbon monoxide and carbon in the form of soot or smoke. The combustion is said to be **incomplete**, and the energy yield from this oxidation reaction will also be lower.

3.3.1.2 Ignition temperature

As we noted in chapter 2, a reaction only proceeds if the appropriate amount of activation energy is supplied to it. A measure of this energy is the resulting rise in the temperature of the reaction mixture. The required temperature is called the **ignition temperature** (or the **ignition point**) for the combustion. It must be noted that sufficient time must be allowed before such a temperature is achieved.

3.3.1.3 Combustion in air and air–fuel ratio

In practice, most combustion is carried out in air rather than in a pure oxygen environment. A method has therefore to be devised to determine the quantity of air that must be used in place of the required quantity of oxygen for complete combustion.

The atmosphere contains 21% oxygen 78% nitrogen and 1% of other gases by volume [3]. This implies that for each mole of oxygen present, there are 3.76 moles of nitrogen and other gases present in the atmospheric gaseous mixture. It is useful to find a way of determining the actual amount of air that will be required to completely burn a given amount of fuel. The concepts of theoretical air and air–fuel ratio are instrumental in this determination.

Definition 3.1: **Theoretical Air** (or **Stoichiometric Air**) is the minimum amount of air required to burn a given amount of fuel completely.

Definition 3.2: **Air–Fuel Ratio (AF)** is the ratio of the mass of air to that of the fuel in a particular combustion process.

The atmospheric composition of gases given above may be used to deduce the **effective molecular mass of air** by finding the weighted average of the molecular masses of oxygen and nitrogen in air. The result obtained is 28.8. This quantity can then be used to evaluate the AF of a given combustion reaction by replacing oxygen with air in the chemical equation for the combustion of a fuel.

In the event that the fuel is natural gas (taken to be 100% methane), the air–fuel ratio is calculated to be 17.14. In practice, excess air is used to ensure complete combustion. Thus, this ratio is chosen to be higher.

(See exercise 3.1—Combustion: Air–Fuel Ratio.)
(See question 3.1—Stoichiometry.)

3.3.1.4 Turbulence

It is intuitively clear that any chemical reaction will only proceed at its theoretically predicted rate if the reacting particles have had every opportunity to interact with each other in the reaction environment. This implies that they are mixed together homogeneously. In practice, this may not be true, and a mechanical means of creating the admixture is needed. A turbulent environment is needed to ensure the required mixing of the reactants.

The requirements for combustion thus include:

- the achievement of the ignition temperature;
- sufficient time to achieve this temperature;
- good turbulence to ensure optimum reaction conditions.

These conditions are often (popularly) stated as:

Time Temperature Turbulence

3.3.2 Combustion of solid biomass

Solid biomass consists of cellular structures constructed from small to large polymeric organic molecules and a fluid interior, and contains a high percentage of water in the intra-cellular and inter-cellular regions. Combustion of biomass is a complex process involving the evolution of its moisture content as well as the degradation of the large biopolymers before the oxidation stage occurs. As a result, the process consists of the following four steps:

- drying;
- devolatilization;
- burning of gaseous volatiles;
- burning of solid char.

These steps occur within different temperature zones. As the temperature rises above ambient, most of the water content is driven off in the region of the biomass called the **drying zone**.

As the temperature increases further, the biomass begins to degrade into smaller volatile compounds (VOCs) in what is called the **devolatilization zone**. Here, the temperature is high enough to induce thermochemical changes in which gaseous products are formed from the thermal breakdown (or decomposition) of the biomass, leaving behind a solid residue called **bio-char**. When the temperature is sufficient for the oxidation of the volatiles, they will burn *in situ*, producing what is called **combustion in the gas phase**. The combustion will appear first as a luminous glow, followed by an incandescent flame as the volatiles begin to burn.

This exothermic reaction may be sufficient to provide the heat to sustain the combustion of the rest of the biomass (i.e. make the combustion **self-sustaining**). At still higher temperatures, the **secondary combustion** of the solid bio-char produced in the devolatilization stage begins to take place.

3.3.3 The products of combustion

When combustion is used as an energy source, the intended output is the **heat (and light) energy** released during the process. This may be used either directly for heating and lighting purposes such as cooking, space heating and lighting, or indirectly through the generation of electrical power for electrical appliances. Carbon dioxide and water are invariably produced as products of reaction when organic matter is burned.

The by-products of combustion occur in solid, liquid and gaseous forms and may be either benign or harmful. The nature of these by-products will depend on the material that is being combusted. They may be generally divided into two major categories [4]:

- solid and liquid wastes such as ash and slag produced by furnaces;
- gaseous and particulate products discharged into the atmosphere.

The solid and liquid waste products of combustion consist of **fly ash**, **bottom ash** and **slag** [5]. Ash is the product of oxidation of the metallic cations contained in the biomass feedstock. It is a powdery substance having a range of particle sizes. The lightest particles form fly ash, which constitutes about half of this total ash content. This component rises to the top of the furnace (boiler) during combustion and will escape into the atmosphere unless a filter mechanism (which may consist of elements such as **scrubbers**, **filters** and **electrostatic precipitators**) exists to prevent its escape.

The heavier component of the ash is called bottom ash, amounting to about 10% of the total ash. This component settles to the bottom of the boiler during combustion and is cleared periodically.

When the temperature exceeds the melting point, ash turns into a molten liquid called slag. In the case of coal or biomass-burning boilers, slag solidifies into a glassy black silica-based solid called 'frit', containing toxic/heavy metals as well as residual carbon char.

Gaseous products discharged in the process include (but are not limited to) **carbon monoxide (CO), nitrogen dioxide (NO_2), carbon dioxide (CO_2), sulphur dioxide (SO_2), water vapour and hydrocarbons (HxCy).**

The particulate matter released into the atmosphere forms **aerosols** (i.e. remain suspended in the air by 'dissolving' in it). These are categorised according to their size as **PM10**, **Coarse Particles** or **PM2.5** [6]. These categories are defined as:

- PM2.5—Fine particles less than or equal to 2.5 μm in diameter.
- Coarse particles—Particles sizes ranging between 2.5 μm and 10 μm.
- PM10—All particles with diameter equal to or less than 10 μm.

Aerosols may be either **primary aerosols** or **secondary aerosols**.

Primary aerosols are suspended particles emitted directly from the combustion process. They consist of **inorganic fly ash**, and products of incomplete combustion.

The latter include **condensable organic compounds (COC), Nitrogen oxides (NOx), Volatile organic compounds (VOCs), Tarry compounds (TAR), polycyclic aromatic hydrocarbons (PAH)** and **Hydrocarbons (HCs).**

Secondary aerosols are formed from the primary aerosols in the atmosphere, and arise from the reaction between the VOCs and other compounds present in the atmosphere. Examples are sulphates formed from SO_2 emission and nitrates formed from NOx emissions. They are divided into **Secondary Organic Aerosols (SOAs)** and **Secondary Inorganic Aerosols (SIAs)** [4].

The degradation or devolatilization of biomass under heat treatment has been mentioned above briefly. This process is formally known as **pyrolysis**. Its two important products are bio-oil, which is of interest because of its use in the production of liquid biofuels for transportation and power generation, and bio-char/charcoal. This process is the subject of more detailed investigation in section 3.4.

(See research question 3.1—Products of combustion: atmospheric pollutants.)

3.4 Pyrolysis

3.4.1 Introduction to pyrolysis

Pyrolysis is the thermal degradation of organic matter (biomass) in the absence of oxygen. This process has been put to productive use by humans since ancient times. The earliest use of pyrolysis (~55 000 years ago) was in the making of charcoal [7] and caulking boats and ingredients for embalming in ancient Egypt. Recent interest in pyrolysis has shifted towards the production of bio-oils [8] and liquid fuels for transportation [9].

The thermal degradation of the biomass results in gaseous volatile products and a solid residue called **bio-char**. The volatile products consist of **condensable** and **non-condensable** (or **permanent) gases**. The condensable products result in the formation of a liquid phase called **bio-oil** or **pyrolysis oil** and **tar** when they are cooled. The net result is the formation of liquid bio-oil, non-condensable gases, tar, solid bio-char and volatile products that are able to escape the reaction environment with the non-condensable gases.

3.4.2 The pyrolysis process

Pyrolysis is used as a dedicated process for the production of products such as charcoal and bio-oil. However, it also forms an integral part of the natural processes of combustion and gasification of solid biomass [10]. We have already noted its role in the combustion of biomass, where it produces the gaseous VOCs and the solid bio-char for the gas and solid phases of combustion. The same two fuels provide the heat energy required for the gasification reactions in the production of syngas (see section 3.5).

Pyrolysis may be effected through the application of heat only or with the assistance of **catalysts**, which speed up the combustion reaction by lowering the activation energy. The thermochemical steps followed in the pyrolysis of biomass in a pyrolysis reactor are similar to those in the combustion or gasification of biomass (with the exception of the combustion reactions). The outputs of the process, however, depend very much on the process parameters in the reaction environment (e.g. the process parameters inside the reactor) and the chemical composition of the feedstock.

In the process of pyrolysis, the substance undergoing pyrolysis (called the fuel or feedstock, e.g. biomass or other carbonaceous solid substance) is chopped and/or ground into smaller size (called **particle size**) and introduced into the reactor which maintains an inert (oxygen-free) environment. The feedstock is then heated steadily (perhaps in steps) from the ambient to a maximum temperature, called the **pyrolysis temperature**, or the **highest treatment temperature (HTT)**. It is maintained at this temperature until the process is complete, following which the gaseous and solid products are quickly removed from the reactor for further treatment [11].

The sequence of events that take place as the temperature rises is as follows:

- Dehydration of the feedstock takes place at the lowest temperatures as it is raised above ambient.
- Pyrolysis (i.e. the thermal degradation of the feedstock) begins as the temperature increases. Hemicellulose begins degrading at temperatures as low as 250 °C, while cellulose depolymerisation commences near 325 °C [12]. The pyrolysis products consist of condensable gases (including tar vapours), non-condensable gases and solid char.
- The condensable products form bio-oils and tars upon condensation, which may occur when contact is made with a cold surface, or if the system includes a dedicated condenser for the condensation and collection of these fractions.
- At still higher temperatures, secondary reactions between the volatiles and tar vapours take place (in the gas phase) in the reaction environment.

The important parameters determining the nature and yields of the products are:

- **particle size;**
- **pyrolysis temperature** (the temperature inside the reactor at which pyrolysis takes place);
- **heating rate** (the time rate of temperature rise in bringing the reactor from the ambient to the pyrolysis temperature); and
- **residence time** (time for which the pyrolysis reactions are allowed to continue).

3.4.3 Types of pyrolysis

Pyrolysis is classified into four broad categories, depending on the temperature, heating rate and residence time. These are:

- very slow pyrolysis (carbonisation);
- slow (conventional) pyrolysis;
- fast pyrolysis;
- flash pyrolysis.

3.4.3.1 Carbonisation

Carbonisation (or very slow pyrolysis) is the thermochemical treatment of solid biomass carried out in oxygen-deficient environments over very long periods of time (days) and very low heating rates. The resulting product is high in carbon content. The process is used in charcoal formation using kilns made from simple pits and mounds to those fabricated from brick, steel or concrete [13].

A striking example of the carbonisation process is contained in the archaeological remains of the city of Pompeii (in modern Italy), which was destroyed by the volcanic eruption of Mount Vesuvius and the resulting lava flow in 79 AD. Recent excavations reveal the preservation of wooden furniture as perfect charcoal replicas of the original objects [14].

3.4.3.2 Slow pyrolysis

Slow pyrolysis is carried out at atmospheric pressure in an inert (oxygen-free) environment. In a typical case [15], the feedstock (which could vary from woody or herbaceous biomass to algae and green waste) is pyrolyzed in a fixed bed reactor consisting of a vertically-arranged steel cylinder which is heated by a furnace and provided with a constant flow of nitrogen to maintain the required inert environment. The feedstock is entered from the top of the reactor. The gaseous products leave the reactor to enter a condenser where the condensable gases are liquified into bio-oil and separated from the permanent gases.

The vapour residence times are long (typically 5–30 min). Low heating rates (<1 °C s^{-1}) are used, and the pyrolysis temperatures are chosen in the 250–700 °C range [16]. The products (bio-oil, gas, bio-char) occur in equal amounts.

3.4.3.3 Fast pyrolysis

The main interest in pyrolysis today is in the production of bio-oils. Higher yields of bio-oils are achieved by increasing the percentage of the bio-oil produced in comparison to bio-char and permanent gases in the products of pyrolysis. In fast pyrolysis, this increase in the bio-oil fraction is achieved by increasing the heating rate so that the pyrolysis temperature is reached before extensive decomposition of the feedstock occurs [11].

The process parameters used in fast pyrolysis to attain this goal include:

- very high heating rates;
- a medium pyrolysis temperature within the 400–600 °C range;
- short residence times (typically 0.5–2 s); and
- rapid quenching of product gas to avoid vapour phase decomposition reactions that reduce the bio-oil fraction [11, 17].

There is currently great interest in the commercial-scale production of fast pyrolysis products [18].

3.4.3.4 Flash pyrolysis

Even higher yields of bio-oil can be obtained through flash pyrolysis (sometimes called very fast pyrolysis). This process is characterised by [19]:

- heating rates exceeding 1000 °C s^{-1};
- high pyrolysis temperatures (in the 900 °C–1300 °C range).

These parameters lead to:
- high bio-oil yields;
- low water content in the products;
- efficiencies exceeding 70%;
- residence times lower than 0.5 s.

Table 3.1 provides a summary of the process parameters and yields obtained by the various types of pyrolysis discussed above.

In chapter 2, **torrefaction** was noted as another thermochemical process applicable to biomass. This is used to improve the energy content of biomass through the processes of drying and thermal densification. In this process, biomass is heated in an oxygen-deficient environment at temperatures within a 280–320 °C range, with residence times ranging from 10 to 30 min. Tumuluru *et al* discuss the various reactors used for biomass torrefaction, and present a model for designing a moving bed torrefier [21].

3.4.4 Products of pyrolysis

The important products of pyrolysis consist of liquid bio-oil (also called pyrolysis oil), solid bio-char and non-condensable gases (including syngas), with some residual water. The relative yields of these depend on the pyrolysis temperature. This temperature dependence is evident in the work carried out by the IEA as reported in their 2006 Annual Report [22]. It reveals that, within the temperature range of 400–650 °C:

- the yield (wt%) of the bio-oil produced rises from 50% at low temperatures to reach a broad peak at ~470 °C, falling gradually as the temperature increases further;
- the bio-char yield falls gradually from ~30% towards values below 10%;
- the yields of the non-condensable (permanent) gases increase from values below 10% towards a maximum of 40%.

The tapering off of the bio-oil yields and the concomitant rise in the non-condensable yields at higher temperatures may be explained heuristically by asserting that the condensable products of pyrolysis (i.e. bio-oil) with larger molecular weights are decomposed into smaller (gaseous, non-condensable) molecules as the temperature increases. This sees an increase in the gaseous products of pyrolysis with temperature. In the case of the solid bio-char, the continuing loss of mass through the devolatilization process results in a progressive reduction in the yield as the temperature rises. These observations are broadly supported by experimental results (see [16] and references therein).

3.4.4.1 Applications of pyrolysis

The major commercial interest in pyrolysis today is in the bio-oil component. This product finds several uses. The applications of pyrolysis can be broadly categorised into [16]:

Table 3.1. Typical process parameters and product yields obtained from different types of pyrolysis (sources: [16, 20]).

Type of pyrolysis	Residence time	Heating rate (°C s^{-1})	Pyrolysis temperature (°C)	Product yield (wt%)		
				Bio-oil	Bio-char	Gas
Very slow (carbonisation)	Several days [20]	Very low [20]	~400 [20]	~0 [20]	~100 [20]	~0 [20]
Slow	5–10 min [16]; 5–30 min [20]	0.1–1 [16]; moderate [20]	280–680 [16]; 600 [20]	30 [16]	35[16]	35 [16]
Fast	0.5–10 s [16]; 0.5–5 s [20]	10–200 [15]; very high [20]	580–980 [16]; 650 [20]	50 [16]	20 [16]	30 [16]
Flash	<0.5 s [16]; <1 s [20]	>1000 [16]; high [20]	780–1030 [16]; <650 [20]	75 [16]	12 [16]	13 [16]

- production of gaseous and liquid fuels including hydrogen, syngas and synthetic fuel products of syngas;
- production of chemicals including resins, fertilisers, food flavours and adhesive; and
- power generation from the direct use of bio-oil in furnaces as well as diesel fuel production through its refinement.

3.4.4.2 Syngas production

The availability of carbon and water in the reactor environment should enable the production of hydrogen and carbon monoxide, the two components of syngas, through carbon gasification, a process that is facilitated under appropriate conditions. It is indeed verified experimentally that the non-condensable phase of the gaseous products produced in pyrolysis consists of a gaseous admixture containing syngas as a component.

Kantarelis and Zabaniotou have carried out the fast pyrolysis of cotton stalk for syngas production [23]. Their results show distinct temperature dependence in the yields of the individual components of the gaseous pyrolysis products. As the temperature increases within the 400–800 °C interval, the percentage of hydrogen in the mixture increases steadily from 20% to 40%, while that of carbon monoxide increases from 40% to 45%. The level of methane present in the gaseous mixture falls gradually from ~15% to values as low as 5%.

3.4.5 Pyrolysis reactors

The essential components of a commercial pyrolysis system used, for instance, to extract pyrolysis oil from waste tyres, rubber and plastic products consists of [24]:

- the feedstock **pre-treatment system** where the feedstock is prepared (e.g. cut and ground) for feeding into the reactor;
- the **pyrolysis reactor** where the pyrolysis of the feedstock is carried out;
- the **cooling system** comprising the **condensers** to quench the volatile gases and **storage tanks** to store the pyrolysis oil;
- the **flue-gas treatment system** to collect and purify the flue gases to the required environmental standards; and
- the **slag discharge system** to remove the slag from the reactor.

If the feedstock is biomass, then:

- the process would begin with the drying of the biomass to moisture contents of less than 10%;
- the biomass would be cut to the chop-length appropriate for the reactor at the pre-treatment stage; and
- the solid discharge would be the bio-char, which has to be removed immediately at the end of pyrolysis to prevent catalysis of the bio-oils.

A variety of reactors are in existence for both scientific research as well as commercial production purposes. They range widely in their size and principles of

Table 3.2. Brief descriptions of some pyrolysis reactors in common use (after [16]).

Reactor	Brief description
1 Fixed bed (also called moving bed) reactor	The reactor is a vertically-placed cylindrical chamber, and the feedstock (fed from the top end of the cylinder) is heated by a furnace at the bottom end. As the feedstock moves down the chamber, the gaseous products of pyrolysis move upwards through it and are collected at the top for condensation and further treatment.
2 Fluidised bed reactors	Here the biomass is introduced as finely-ground particles which are mixed with a heated particulate substrate for rapid exchange of heat. In a **Bubbling Fluidised Bed** reactor, the solid substrate is heated sand, and the biomass decomposes into gas, vapour, aerosol and char. The **Circulating Fluidised Bed** reactor works on the same principle but has shorter residence times for the vapours and char.
3 Ablative reactors	These consist of the **ablative vortex reactor** and the **ablative rotating disk reactor**. They produce pyrolysis through mechanical pressure applied between the solid biomass and a heated reactor wall. In the former, biomass particles are carried in a hot inert gas that enters tangentially in the vortex reactor. In the rotating disk reactor, the biomass is pressed against the rotating disk to effect the pyrolysis reactions.
4 Vacuum pyrolysis reactor	Here, biomass is carried into a vacuum chamber held at a high temperature via a moving belt and heated using burners and induction heaters.
5 Rotating cone reactor	Here, hot sand and biomass are mixed mechanically in a rotating cone-shaped reactor at its base and move to the top through centrifugal forces, releasing pyrolysis vapours to a condenser.
6 PyRos reactor	Here, the biomass and the inert particulate matter are introduced into a cyclone and move to the periphery through centrifugal forces.

operation. Table 3.2 provides a brief introduction to six most prominent pyrolysis reactors in use today.

(See research question 3.2—The fixed bed and ablative pyrolysis reactors.)

(See quiz 3.1—Pyrolysis.)

3.5 Gasification

3.5.1 What is gasification?

As indicated in section 3.2.2, the nature of a thermochemical process depends on the process parameters. Examples of these include temperature, pressure, residence time and the oxygen and water content of the reaction environment (which may be provided by the conditions within a reacting chamber).

When oxygen is completely absent, the thermochemical treatment of biomass leads to its pyrolysis. When it is available in the stoichiometric quantity or in excess, complete combustion occurs, with carbon dioxide and water forming the primary products of the process.

Gasification occurs when the oxygen content is insufficient for stoichiometric combustion. It is partial oxidation (i.e. less oxygen is used than that required for the complete combustion of the fuel). Gasification typically uses 25–40% of the stoichiometric quantity required for complete oxidation. Water (in the form of steam) is used in the process to facilitate the production of products such as hydrogen and carbon monoxide from the carbonaceous material present, as well as a means of moderating the temperature.

The main product of gasification is a gaseous mixture, called variously as **syngas** or **producer gas**, that contains varying amounts of **hydrogen** and **carbon monoxide.** By-products consist of gases such as carbon dioxide and smaller quantities of (possibly toxic) gases, as well as solid or liquid ash and/or slag.

3.5.2 Overview of the gasification process

Gasification is a complex process that includes the physical as well as chemical transformation of carbonaceous feedstock into syngas together with by-products. The stages that take place in gasification are, in principle, similar to the temperature-dependent steps of combustion described in section 3.2. All three processes of pyrolysis, combustion and gasification occur within a gasification reactor when the feedstock consists of coal, biomass or biomass-derived waste.

3.5.2.1 The chemical changes [25]

The chemical changes that take place in a dedicated gasification reactor are:

- devolatilization of feedstock to produce volatile products such as tar vapours, hydrogen and methane, as well as solid bio-char;
- combustion of volatiles and some bio-char with the limited supply of oxygen to produce carbon dioxide, carbon monoxide and the release of energy for the gasification reaction that follows;
- reaction of the remainder of char with carbon dioxide and water (steam) to produce carbon monoxide and hydrogen (the two components of syngas).

In addition, there are two reactions that take place in the gas phase, namely the **water–gas shift reaction** and the **steam–methane reforming reaction** that enhance the hydrogen content of the syngas mixture.

3.5.2.2 The end products

The end result of gasification is the production of syngas, with small amounts of water and carbon dioxide, methane and other gases (the relative amounts depending on the selected process parameters), as well as ash or slag together with particulate atmospheric pollutants.

Table 3.3 describes the composition of the gaseous products resulting from a typical gasification process.

Table 3.3. The composition (vol%) of the gaseous products of a typical gasification process in a coal-fired gasifier. Data source: Department of Energy, National Energy Technology Laboratory. Data source: [25].

Gas	Relative amount in final product (vol%)
H_2	25–30
CO	30–60
CO_2	5–15
H_2O	2–30
CH_4	0–5
H_2S	0.2–1
COS	0–0.1
N_2	0.5–4
Ar	0.2–1
NH_3 and HCN	0–0.3

3.5.2.3 The temperature zones of gasification

The typical stages in gasification are most obvious in the fixed (or moving) bed gasifier to be described later. Box 3.1 provides a summary of these stages in terms of the temperature zones of gasification [25]:

(See exercise 3.2—Comparison of temperature zones during combustion and gasification.)

To obtain a fuller picture of the process, one needs to elaborate further on the possible chemical reactions that take place. This is most readily done in the case where the feedstock is a highly carbonaceous solid such as coal.

3.5.3 The chemistry of gasification

In the gasification of coal, the feedstock (coal) is predominantly carbon in its elemental composition. The reactions that occur during gasification fall into three categories. These consist of combustion reactions releasing the heat required for the gasification, reactions involving steam, and two special reactions that assist in enhancing the hydrogen content of syngas [25].

Combustion reactions between the devolatilization products (char, carbon monoxide and hydrogen) and oxygen:

$$C + \tfrac{1}{2}O_2 \rightarrow CO \; \Delta E = (-111 \text{ MJ/kmol}^{-1}) \tag{3.2}$$

$$CO + \tfrac{1}{2}O_2 \rightarrow CO_2 \; \Delta E = (-283 \text{ MJ/kmol}^{-1}) \tag{3.3}$$

$$H_2 + \tfrac{1}{2}O_2 \rightarrow H_2O \; \Delta E = (-242 \text{ MJ/kmol}^{-1}) \tag{3.4}$$

Reactions involving steam, carbon and the products of previous reactions:

$$C + H_2O \leftrightarrow CO + H_2 \text{ (the Water–Gas Reaction, } \Delta E = +131\ \text{MJ/kmol}^{-1}\text{)} \quad (3.5)$$

$$C + CO_2 \leftrightarrow 2CO \text{ (the Boudouard Reaction, } \Delta E = +172\ \text{MJ/kmol}^{-1}\text{)} \quad (3.6)$$

$$C + 2H_2 \leftrightarrow CH_4 \text{ (the Methanation Reaction, } \Delta E = -75\ \text{MJ/kmol}^{-1}\text{)} \quad (3.7)$$

Under special conditions, the above reactions may be replaced by two reactions that occur in the gas phase, and that increase the relative abundance of hydrogen in the gasification process. These are the **water–gas shift reaction**

$$CO + H_2O \rightarrow CO_2 + H_2 \quad \Delta E = (-41\ \text{MJ/kmol}^{-1}) \quad (3.8)$$

And the **steam–methane reforming reaction**

$$CH_4 + H_2O \rightarrow CO_2 + 3H_2 \quad \Delta E = (+206\ \text{MJ/kmol}^{-1}) \quad (3.9)$$

Box 3.1. The temperature zones of gasification.

The gasification process may be viewed in terms of the sequence of the four temperature regimes or zones of **drying, devolatilization (pyrolysis), combustion** and **gasification**.

Drying

At the lowest temperatures above ambient, drying or **dehydration** of the feedstock takes place in which any free water in the feedstock is driven off.

Pyrolysis

Pyrolysis occurs as the temperature increases, during which the solid devolatilizes through the breaking of weaker chemical bonds, resulting in the release of volatile tar vapours, methane and hydrogen, while leaving behind a solid carbonaceous residue called **char**. More specifically, when the feedstock is coal, devolatilization begins above 400 °C, with the products consisting of char, **condensable liquids** and **light gases** [26].

In the case of a biomass feedstock, this stage of gasification involves thermal degradation (cracking) reactions where solid char and light gases such as H_2, CO, CO_2, CH_4, H_2O, NH_3 and tarry particles are released. Part of the vapours produced also undergoes thermal cracking to gas and char. For this feedstock, volatile matter always represents a significant portion of the resulting products [27].

In the case of unconventional feedstocks such as heavy oil and tar sands, devolatilization starts within the lower temperature range of 200–300 °C. The volatiles consist of a range of products varying from carbon monoxide and methane to high molecular weight hydrocarbons including alkanes, alkenes, aromatic hydrocarbons, heavy oil and tar [28].

Combustion

Both products of pyrolysis (i.e. the volatile gases as well as some of the solid char) undergo combustion in the limited oxygen supply available to produce carbon dioxide and carbon monoxide. The heat produced in this exothermic reaction is used as the energy input into the endothermic gasification reactions that follow.

Gasification

At temperatures above ~600 °C, the remaining char reacts with carbon dioxide and the added steam in the gasification reactions that produce the carbon monoxide and hydrogen that comprise syngas [26].

3.5.4 Uses of syngas

3.5.4.1 Overview

Syngas, the main product of the gasification process, has many uses. This gaseous mixture finds use both as a direct fuel as well as providing an intermediate product in the product chain of several manufacturing processes. Syngas may be used as a fuel for power plants, or in the production of secondary fuels for the transportation industry and other uses through the **Fischer–Tropsch Synthesis (FTS)** process.

More specifically, syngas may be used:

- as the gaseous fuel to power gas turbines in **Integrated Gasification Combined Cycle Power Plants (IGCCPP)**;
- in the production of transportation and power generation fuels and other substances through the Fischer–Tropsch Synthesis process [29];
- to provide the hydrogen needed in the production of ammonia and nitrogen-based fertilisers;
- to provide 'blue' hydrogen for fuel cells to generate electricity as well as an alternative fuel source for domestic cooking and heating needs for the future.

Syngas provides the fuel used for the generation of electricity in Integrated Gasification Combined Cycle Power Plants. These installations produce electricity by feeding syngas produced by a gasifier directly into the gas turbine of the power plant.

The syngas produced in the steam–methane reforming process has been traditionally used as the source of hydrogen for the production of ammonia gas. This, in turn, is used to produce nitrate-based fertilisers.

More recently, there has been great interest shown in the USA and Australia in the production of blue hydrogen (named as its production through coal gasification involves the release of the greenhouse gas carbon dioxide) as part of their emissions reduction strategies [25, 30].

3.5.4.2 Fischer–Tropsch synthesis

Historically, perhaps one of the most significant uses of syngas has been as the feedstock in the Fischer–Tropsch Synthesis (FTS) process. The process can be used to produce synthetic hydrocarbon fuels and various chemicals from a range of carbonaceous feedstock that produces syngas as an intermediate product. This process will be discussed further in chapter 11.

3.5.5 Gasifiers

Reactors used in the gasification of coal and biomass are called **gasifiers**. Many designs exist, but they fall into the following three main classifications:

- **Fixed bed gasifiers** (also called **moving bed gasifiers**)—where the feedstock (in the form of large particles) moving down the gasifier reacts with the high-oxygen content gas and steam moving up the reactor. These operate at moderate pressures, and reactions within the gasifier occur within different temperature zones as described above. Examples are the Lurgi Dry Ash Gasifier and the British Gas/Lurgi (BGL) gasifier.
- **Entrained flow gasifiers**—where the finely-ground feedstock is fed concurrently with oxygen and steam from the top of the reactor. They are called 'entrained flow' as the air and steam surround or 'entrain' the feedstock particles as they move through the gasifier in a cloud.
- **Fluidised bed gasifiers**—where the fine feedstock particles are **suspended** in an oxygen-rich gas and steam mixture fed from the bottom of the reactor. The bed acts as a fluid. Particle size must be small (<6 mm) for fluidisation to be accomplished.

Commercial examples of the fixed bed gasifier are the **Lurgi** and the **British Gas Lurgi (BGL)** reactors. Examples of entrained flow gasifiers are provided by the **GE Energy**, **CB&I E-Gas** and **Shell SCGP** gasifiers. Examples of fluidised bed gasifiers include the **Great Point Energy**, **Winkler** and the **KBR Transport** gasifiers.

3.6 Summary of thermochemical processes and products

The above sections have provided an overview of the thermal and thermochemical processes of combustion, pyrolysis and gasification with a view to discussing their science and industrial significance. It is obvious from the above discussions that these three processes often form part of the same overall process, and therefore there is a need to differentiate them clearly. This can be partly achieved by showing how the three processes are related and enumerating the intended end product of each. Table 3.4 attempts to do this by listing the end products and enumerating the interdependencies amongst these three processes.

Table 3.4. Thermochemical processes and their products.

Process	Desired end product/s	Auxiliary processes required	All products and by-products
Combustion	Heat (and light) energy	Drying, devolatilization of feedstock	Heat, water, carbon dioxide; Ash, slag, atmospheric pollutants
Pyrolysis	Bio-oils, bio-char, syngas	Drying, combustion of volatiles	Bio-oil, charcoal, syngas; water

(*Continued*)

Table 3.4. (*Continued*)

Process	Desired end product/s	Auxiliary processes required	All products and by-products
Torrefaction	Densified feedstock	None in particular	Densified biomass
Gasification	Syngas	Drying, devolatilization of feedstock, combustion of volatiles and char	Syngas; ash, slag, atmospheric pollutants

Exercises, questions and quizzes

Exercises

Exercise 3.1. Combustion: Air–Fuel Ratio.

Obtain the value of the effective molecular mass of air by carrying out the suggested mathematical exercise in section 3.2.1, and verify the result for the AF for methane.

Exercise 3.2. Comparison of temperature zones during combustion and gasification.

Compare the temperature zones for gasification enumerated in box 3.1 with those met in section 3.2.2 for combustion, and identify each step as belonging to either combustion, pyrolysis or gasification.

Questions

Question 3.1. Stoichiometry.

What mass of air would be required for the complete combustion of one mole of butane gas? Comment on the assumptions that you have made in arriving at this estimate.

Research questions (RQs)

Research question 3.1. Products of combustion: atmospheric pollutants.

a) Carry out a desktop investigation into the types of atmospheric pollutants produced in the combustion of fossil fuels, and the incineration of municipal wastes. Many of these are harmful to human health, including some which are carcinogenic.
b) Describe the health hazards posed by the two classes of toxic compounds known as **dioxins** and **furans**, and briefly discuss their molecular structures. Identify at least two national and/or global authorities responsible for the monitoring and regulation of such atmospheric pollutants.

Research question 3.2. The fixed bed and ablative pyrolysis reactors.

a) Describe, with the help of detailed schematic diagrams, the principle of operation of the **fixed Bed** and **ablative** pyrolysis reactors. The diagrams must include the furnace and all components for the condensation and storage of the products.

b) Explain the need for the pre-treatment of woody biomass feedstocks, and describe how this is carried out.

Quizzes

Quiz 3.1. Pyrolysis.

a) What is another term used to describe the thermal degradation of biomass?
b) Name an essential condition necessary for the successful pyrolysis of biomass, and list the type of solid, liquid and gaseous products formed.
c) Define the terms:
   (i) particle size;
   (ii) heating rate;
   (iii) pyrolysis temperature.
d) Describe the main differences between slow and fast pyrolysis, and comment on the types and yields of the products formed in each.
e) Name the essential components of a commercial pyrolysis system.

References

[1] Phlogiston chemical theory. Britannica [accessed 4 November 2022]. Available from https://britannica.com/science/phlogiston

[2] History of the study of combustion. Britannica [accessed 5 November 2022]. Available from https://britannica.com/science/combustion/History-of-the-study-of-combustion

[3] Earth Now. The Atmosphere. NASA Global Climate Change [accessed 4 November 2022]. Available from https://climate.nasa.gov/news/2915/the-atmosphere-getting-a-handle-on-carbon-dioxide/

[4] Solid biomass combustion plants and by-products. Technosida [accessed 21 November 20210]. Available from https://tecnosida.com/biomass-combustion-plants-and-by-products

[5] Coal ash, fly ash, bottom ash and boiler slag: a top to bottom tour of fossil fuel sludge. Natural Resources Defence Council (NRDC) [accessed 5 November 2022]. Available from https://nrdc.org/onearth/coal-ash-fly-ash-bottom-ash-and-boiler-slag

[6] Particulate matter (PM) basics. US EPA [accessed 5 November 2022]. Available from https://epa.gov/pm-pollution/particulate-matter-pm-basics

[7] Antal M J and Gronli M 2003 The art, science and technology of charcoal production *Ind. Eng. Chem. Res.* **42** 1619–40

[8] Demirbas A 2002 Partly chemical analysis of liquid fraction of flash pyrolysis products from biomass in the presence of sodium carbonate *Energy Convers. Manag.* **43** 1801–9

[9] Demirbas A 2004 Combustion characteristics of different biomass fuels *Prog. Energy Combus. Sci.* **30** 219–30

[10] Grønli M G, Varhegyi G and Blassi C D 2002 Thermogravimetric analysis and devolatilization kinetics of wood *Ind. Eng. Chem. Res.* **41** 4201–8

[11] Basu P 2018 *Biomass Gasification, Pyrolysis and Torrefaction – Practical Design and Theory* 3rd edn (Melbourne: Elsevier)

[12] Yang W, Ponzio A, Lucas C and Blasiak W 2006 Performance analysis of a fixed-bed biomass gasifier using high-temperature air *Fuel Proc. Technol.* **87** 235–45

[13] Antal M J and Gronli M 2003 The art, science, and technology of charcoal production *Ind. Eng. Chem. Res.* **42** 1619–40

[14] Snow D and Rashka D 2021 Pompeii: After the Eruption. SBS TV Melbourne, Australia

[15] Ronsse F, van Hecke S, Dickinson D and Prins W 2013 Production and characterization of slow pyrolysis biochar: influence of feedstock type and pyrolysis conditions *Glob. Change Biol. Bioenergy* **5** 104–15

[16] Jahirul M I, Rasul M G, Chowdhury A A and Ashwath N 2012 Biofuels production through biomass pyrolysis – a technical review *Energies* **5** 4952–5001

[17] Luque R, Lin C S K, Wilson K and Clark J (ed) 2016 *Handbook of Biofuels Production – Processes and Technologies* 2nd edn (Amsterdam: Elsevier) p 748 Available from https://sciencedirect.com/book/9780081004555/handbook-of-biofuels-production

[18] Biomass pyrolysis research at the Eastern Regional Research Centre. USDA Agricultural Research Service; October 2021 [accessed 5 November 2022]. Available from https://ars.usda.gov/northeast-area/wyndmoor-pa/eastern-regional-research-center/docs/biomass-pyrolysis-research-1/what-is-pyrolysis/

[19] Suib S L (ed) 2013 *New and Future Developments in Catalysis: Catalysis by Nanoparticles* (Amsterdam: Elsevier)

[20] Mohan D, Pittman C U and Steele P H 2006 Pyrolysis of wood/biomass for bio-oil: a critical review *Energy Fuels* **20** 848–89

[21] Tumuluru J S, Ghiasi B, Soelberg N R and Sokhansanj S 2021 Biomass torrefaction process, product properties, reactor types and moving bed reactor design concepts *Front. Energy Res.* **9** 728140

[22] International Energy Agency 2006 *Annual Report, 2006: IEA Bioenergy: Task 34, Pyrolysis of Biomass* (Paris: International Energy Agency)

[23] Kantarelis E and Zabaniotou A 2009 Valorization of cotton stalks by fast pyrolysis and fixed bed air gasification for syngas production as precursor of second generation biofuels and sustainable agriculture *Bioresour. Technol.* **100** 942–7

[24] Components of pyrolysis plant. Beston Group 2021 [accessed 6 November 2022]. Available from https://bestongroup.com/components-of-pyrolysis-plant/

[25] Gasification Introduction. National Energy Technology Laboratory [accessed 6 November 2022]. Available from https://netl.doe.gov/research/Coal/energy-systems/gasification/gasifipedia/intro-to-gasification

[26] Luque R and Speight J G 2015 *Gasification for Synthetic Fuel Production: Fundamentals, Processes and Applications* (Sawston: Woodhead Publishing) p 333

[27] Arena U 2013 ed F Scala *Fluidized Bed Technologies for Near-Zero Emission Combustion and Gasification* (Sawston: Woodhead Publishing)

[28] Speight J G 2014 *Gasification of Unconventional Feedstocks* ed J G Speight (Houston, TX: Gulf Professional Publishing)

[29] Van de Loosedrecht J, Botes F G, Ciobica I M, Ferreira A, Gibson P and Moodley D J *et al* 2013 Fischer–Tropsch synthesis: catalysis and chemistry ed J Reeddijk and K Poepellmeier *Comprehensive Inorganic Chemistry II* vol 7 (Oxford: Elsevier) ch 7.20 pp 525–57

[30] COAG Energy Council 2019 Australia's National Hydrogen Strategy. Australian Government Department of Climate Change, Energy, Environment and Water [accessed 6 November 2022]. Available from https://dcceew.gov.au/energy/publications/australias-national-hydrogen-strategy

IOP Publishing

Bioenergy for Power Generation, Transportation and Climate Change Mitigation

Anirudh Singh

Chapter 4

Chemical, biochemical and electrochemical conversion of biomass

Chemical, biochemical and electrochemical processes play significant roles in the production of biofuels. This chapter provides a first introduction to the basic science of the production of biodiesel, bioethanol and biogas, and briefly discusses hydrogen and hydrogen fuel cells.

The chapter begins by discussing the conventional and emerging chemical methods used in the conversion of triglycerides found in plant and animal biomass into biodiesel and hydro-treated vegetable oil (HVO), and briefly mentions the chemical synthesis of transportation fuels from syngas using the Fischer–Tropsch Synthesis (FTS) process. The biochemical processes of anaerobic respiration and methane fermentation used in the production of bioethanol and biogas are then introduced by first providing a detailed introduction to metabolic pathways and the role of enzymes in biochemical processes. The chapter ends with a brief reference to hydrogen as a chemical energy storage medium and the role of hydrogen fuel cells in converting chemical energy into electricity.

4.1 Introduction

Chapter 3 considered the thermal and thermochemical methods of converting the energy content of biomass into a form appropriate for end use. This chapter completes the introduction of the various conversion pathways introduced in figure 3.1 by considering the chemical, biochemical and electrochemical pathways available for biomass conversion. It introduces the scientific principles that provide the basis for further discussion of liquid and gaseous biofuels considered in parts II and III of the book.

Section 4.2 first introduces the energy-rich compounds known as **lipids**, and shows how the simple lipids, consisting of naturally-occurring fats, oils and waxes, can be

doi:10.1088/978-0-7503-3555-3ch4

converted to mixtures of esters called **biodiesel** through the process of **transesterification**. It next demonstrates how these same organic feedstocks can be converted into more engine-compatible fuels called **Hydro-treated Vegetable Oil (HVO)**. The chemical conversion of syngas into synthetic fuels for transportation is mentioned briefly.

The biochemical conversion of biomass into bioethanol and biogas is discussed in section 4.3 by first introducing the chemical processes involved in cellular metabolism. The section begins by investigating the basic chemistry of catalysis, and points out that biochemical reactions are different from chemical reactions in that each biochemical reaction is assisted by its own biological catalyst called an enzyme, and that such reactions occur in chains known as metabolic pathways. Fuels such as bioethanol are produced by the biochemical process of anaerobic respiration.

The next section introduces the process of anaerobic digestion in which consortia of bacteria participate to convert the polymeric content of biomass into biogas through the process of methane fermentation. Section 4.5 is a brief introduction to the direct conversion of the chemical energy of biomass into electricity using fuel cells.

4.2 Chemical conversion of biomass

4.2.1 Transesterification and biodiesel

4.2.1.1 Some biochemistry background

Lipids, including fats and oils, are a whole class of plant and animal-derived bioenergy resources that have been available as fuel for human consumption since antiquity. These compounds have in recent times begun to play important roles in the search for environmentally friendly alternative fuels for transportation. Before examining the conversion of this type of feedstock for the fuel industry, it will be instructive to look briefly at their biological origins.

Lipids are a large group of organic compounds, including fats, oils and hormones, that are generally insoluble in water, but form an integral part of the biochemistry of living organisms [1]. Some lipids are amphipathic, i.e. have two ends, one of which is soluble in water (i.e. is hydrophilic) and the other is not (i.e. is hydrophobic). Such lipids play important roles in the formation of cell and nuclear membranes of all cellular organisms [1].

Lipids can be classified as **simple or complex lipids**. The simple lipids consist of the **fats, oils and waxes** occurring in nature, while complex lipids include **phospholipids** and **glycolipids**. Our interest is in the simple lipids found in plants and animal biomass. These are high energy-content organic polymers (see next section) and thus provide an important class of feedstock for the production of biofuels for the transportation industry. They are naturally-occurring **triglycerides**, which can be transformed into alternative fuels for diesel engines.

4.2.1.2 Biodiesel as transportation fuel

As was known to Rudolph Diesel himself, Diesel engines can run on vegetable oil, sometimes called pure plant oil (PPO) [2]. Such triglycerides, however, suffer from the disadvantage that they have very high viscosities. One of the essential fuel properties of diesel engine fuels is that they should have low viscosities. This is necessary for their ease of atomisation during the injection stages of the diesel engine power cycle (see part III for more discussion of this topic).

(See question 4.1—Properties of diesel engine fuels.)

4.2.1.3 Transesterification of triglycerides and biodiesel production

A triglyceride is an **ester** that is formed when a fatty acid reacts with glycerol, an alcohol with three hydroxyl groups (i.e. a triol). Fatty acids can have very long hydrocarbon chains, making the molecular weight of the ester very large. Triglycerides thus have very high viscosities due to their large molecular size, and a way must be found to reduce this viscosity to improve fuel characteristics. This can be achieved through **transesterification**, in which one triglyceride molecule is converted into three smaller ester molecules.

Before seeing how this is achieved, one must know what an ester is and how it is produced. An ester is an organic compound formed by reacting an organic acid (e.g. **carboxylic acid**) with an alcohol. The general formula for a carboxylic acid is *R–COOH*, where *R* is the hydrocarbon chain and the rest of the molecule is called the **carboxylic functional group**.

An example of a carboxylic acid is acetic acid, where the hydrocarbon chain has only a single carbon atom. **Fatty acids are carboxylic acids with generally very large numbers of carbon atoms in the *R* group**. An example of a reaction between a carboxylic acid and alcohol (called an **esterification reaction**) is shown in figure 4.1, where acetic acid reacts with methanol to produce the ester called methyl acetate and a molecule of water.

In a **transesterification reaction**, an existing ester (a triglyceride) is reacted with an alcohol such as methanol in the presence of a catalyst (e.g. NaOH or KOH) to produce new esters and glycerol as products [3]. This mixture of low molecular weight esters is known as **biodiesel**. Figure 4.2 shows a generic triglyceride (with three fatty acid chains denoted R_1, R_2, R_3) reacting with methanol to produce a mixture of low molecular weight esters and the glycerol molecule that had provided the backbone for the triglyceride.

If the alcohol is methanol as in figure 4.2, then the biodiesel that is produced is a mixture of **Fatty Acid Methyl Esters** (abbreviated as **FAME**).

The *R* groups can consist of either **saturated hydrocarbon chains** (i.e. no double bonds between carbon atoms) or **unsaturated chains** (containing one or more double

$$CH_3-\overset{\overset{\displaystyle O}{\|}}{C}-OH + CH_3OH \rightarrow CH_3-\overset{\overset{\displaystyle O}{\|}}{C}-CH_3 + H_2O$$

Acetic Acid   Methanol   Methyl Acetate   Water

Figure 4.1. An example of an esterification reaction, where acetic acid reacts with methanol to produce the ester called methyl acetate and a molecule of water.

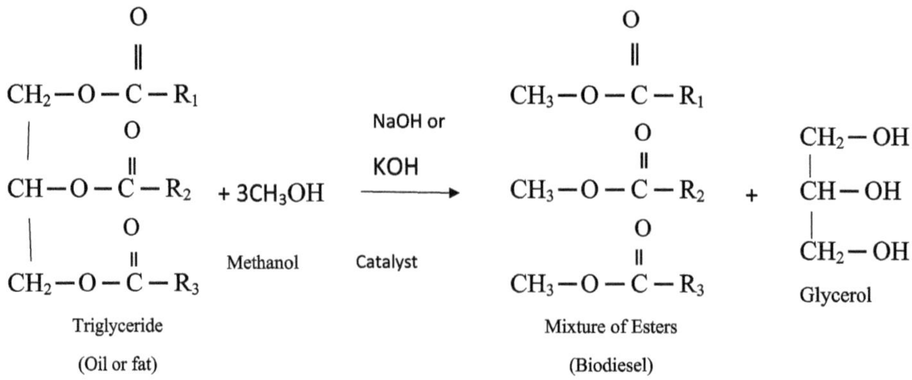

Figure 4.2. The transesterification process in the production of biodiesel. A triglyceride with three fatty acid chains R_1, R_2, R_3 reacts with an alcohol (methanol in this case) in the presence of a catalyst (NaOH or KOH) to produce a mixture of three esters (the biodiesel) and glycerol.

Table 4.1. Composition (percentage content of fatty acid chains) of common naturally-occurring or synthetic oils, fats and consumer products. The first row refers to the number of carbon atoms and the degree of saturation of the fatty acid chain (see text). Data source: [3].

	Triglyceride	14:0	16:0	18:0	18:1	18:2	18:3
1	Soybean		6%–10%	2%–5%	20%–30%	50%–60%	5%–11%
2	Corn	1%–2%	8%–12%	2%–5%	19%–49%	34%–62%	~0%
3	Peanut		8%–9%	2%–3%	50%–65%	20%–30%	
4	Olive		9%–10%	2%–3%	73%–74%	10%–12%	~0%
5	Cottonseed	0%–2%	20%–25%	1%–2%	23%–35%	40%–50%	~0%
6	Hi oleic safflower		5.9%	1.5%	8.8%	83.8%	
7	Butter	7%–10%	24%–26%	10%–13%	28%–31%	1%–2.5%	0.2%–0.5%
8	Lard	1%–2%	28%–30%	12%–18%	40%–50%	7%–13%	0%–1%
9	Tallow	3%–6%	24%–32%	20%–25%	37%–43%	2%–3%	
10	Linseed		4%–7%	2%–4%	25%–40%	35%–40%	25%–60%
11	Yellow grease	2.4%	23.2%	13.0%	44.3%	7.0%	0.7%

bonds). Table 4.1 shows the percentage of different fatty acid chains (*R*) present in common fats, oils and consumer products. The first row of the table indicates the number of carbon atoms in the hydrocarbon chain (which includes the C atom double-bonded to the oxygen atom), separated by a colon from the number of double bonds present in the chain (i.e. degree of unsaturation).

(See exercise 4.1—Calculating biodiesel yield in transesterification.)

4.2.2 Hydro-treated vegetable oils as fuels for transportation

Biodiesel such as a Fatty Acid Methyl Ester (FAME) suffer from several issues with regard to their suitability as green transportation fuels. Apart from the ongoing

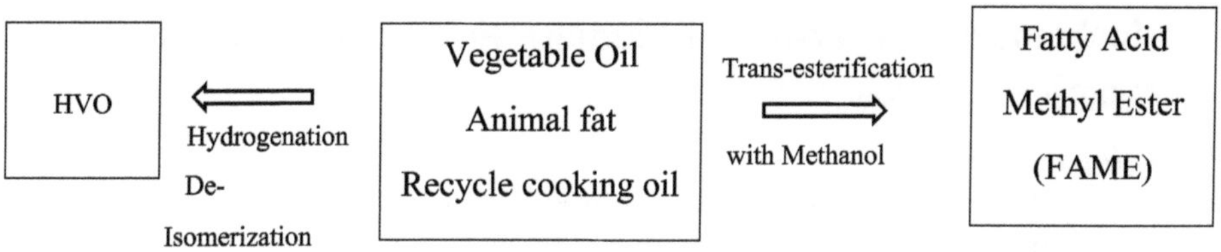

Figure 4.3. Alternative pathways to diesel engine fuel production from vegetable oils: FAME vs HVO.

debate about their true levels of carbon neutrality, and the economics and the ethical concerns (i.e. food versus fuel) relating to FAME production and use [4], other issues include:

- the presence of free methanol, glycerol and fatty acids as impurities due to its production process;
- water contamination; and
- susceptibility to oxidation and biodegradation.

A method of reducing the susceptibility of FAME to oxidation and biodegradation involves the removal of oxygen from its molecular structure and the saturation of double bonds in the hydrocarbon chains. As will be discussed more fully in part III, such a step will also have the added advantage of bringing the fuel closer in molecular structure to petroleum diesel (which consists predominantly of straight and branched-chain hydrocarbons), thus making it easier to satisfy the fuel standards [5].

Hydro-treated vegetable oil (HVO) is a fuel prepared from fats and oils using such chemical treatments. The production of HVO from triglycerides proceeds via the processes of **hydrogenation**, **de-oxygenation** and **isomerisation**, and provides an alternative pathway to diesel engine fuel production from vegetable oils [6]. These alternative pathways are displayed in figure 4.3.

The production of HVO as a green fuel for diesel engines will be discussed more fully in part III.

4.2.3 Synthetic fuels from syngas—the FTS process

We saw in chapter 3 that the main end product of the gasification process was syngas. The hydrogen and carbon monoxide gases that make up this product provide the starting materials for the synthesis of larger hydrocarbon molecules.

The bulk of transportation vehicles today are driven either by internal combustion engines or jet engines, both of which have been designed to be powered by fossil fuels. The three main categories of fuels for transportation are **petroleum diesel** for diesel engines, **petrol or gasoline** for petrol engines and **jet fuel** for air transportation. All these fuels are hydrocarbons obtained through the fractional distillation of mineral (i.e. fossil fuel) oil.

Such hydrocarbon-based fuels can also be synthesised via the **Fischer–Tropsch Synthesis (FTS) process** from syngas derived from the gasification of biomass as well as other sources [7]. A detailed discussion of the FTS process is provided in part III.

4.3 Biochemical conversion of biomass through respiration

4.3.1 Cellular metabolism

4.3.1.1 Introduction

Bioethanol, bio-butanol and biogas are three current or emerging biofuels for the transportation sector. All three are produced through biochemical processes and will be discussed in detail in part III. To fully understand the mechanism of their production however, one requires a thorough understanding of the biochemical processes by which plants, animals and bacteria acquire the energy and material requirements for their sustenance. This process is called **respiration**, and the above biofuels are by-products of the metabolic pathways that provide the chemical basis for respiration.

This section provides a first introduction to the metabolic pathways of aerobic and anaerobic respiration used by organisms that convert bioenergy into **adenosine triphosphate (ATP)** and produce bio-alcohols as by-products. It begins by first examining the differences between the (inorganic) chemical reactions generally encountered in the chemical laboratory and the biochemical reactions that provide the basis for the metabolic pathways of respiration.

4.3.1.2 Difference between chemical and biochemical reactions

The biochemical processes that occur in the cells of organisms are far more complex than the simple reactions encountered in inorganic chemistry. While simple (inorganic) reactions describe the conversion of the reactants to the products in terms of simple chemical equations, biochemical reactions usually consist of a multitude of such reactions that work together to produce the desired end product. These are invariably assisted by a special class of molecules called enzymes.

The reactions within the cellular organism need to be confined within a moderate temperature range close to the ambient temperature at the Earth's surface. It is speculated that this requirement originates from the aquatic environment in which life originally evolved on Earth [8]. As water remains in the liquid phase only between 0 °C and 100 °C, and as (according to the theory of evolution) land-based organisms evolved from the sea, this provides the overall limiting temperature boundaries for the survival of nearly all organisms [8]. One therefore expects that the molecular content of most organisms will be stable within a similar (limited) temperature range.

According to the Arrhenius theory, the rates of chemical reactions (i.e. the rate at which reactants are used up or products formed) depend on both the temperature and the activation energy for the reaction. This is seen by considering the chemical reaction

$$aA + bB \rightarrow cC + dD \tag{4.1}$$

The rate of this reaction is given by

$$\text{Rate} = k[A]^a[B]^b \tag{4.2}$$

where rate constant k is given by the expression

$$k = A\exp[-E_a/RT] \tag{4.3}$$

Here, E_a is the activation energy, T is the absolute temperature and R is the gas constant. (Note that k is a constant only if E_a and T are both held fixed.)

We see that the reaction will proceed faster if

(i) the temperature T is increased, or
(ii) the activation energy E_a is decreased.

Thus, we need high temperatures and/or small activation energies to produce the end products of metabolism at reasonable rates. In inorganic chemistry, the high temperature requirement can be avoided by reducing the activation energy E_a through the use of **catalysts**.

Enzymes are organic molecules that serve the same purpose as catalysts in living organisms, where organic products need to be produced at low temperatures, but at rates which are high enough to sustain life.

In summary, biochemical reactions usually consist of many reactions, linked together in chains, that lead to the final desired products. These chains of enzyme-assisted reactions are called **metabolic pathways**. Each reaction is facilitated by an enzyme, which is frequently produced by micro-organisms.

Table 4.2 summarises the special features and requirements of biochemical reactions as compared to those of simple (inorganic) chemical reactions.

4.3.1.3 Some phylogeny

As noted above, biochemical reactions often involve micro-organisms that assist the process by providing the required enzymes. These organisms are frequently referred to by their phylogenetic classification [9]. It is thus important to gain some basic familiarity with the terminology.

Table 4.2. Special features and requirements of biochemical reactions as compared to inorganic chemical reactions.

Biochemical reactions	Chemical reactions
1. Must occur at moderate temperatures and pressures that are sustainable within the cellular environment.	Can occur at virtually any temperature achievable technologically. In general, high temperatures produce greater reaction rates.
2. Require the presence of biological catalysts, called enzymes, to facilitate the production of the end product.	Catalysts are not always required. However, inorganic catalysts are often used to increase product production rate and yields.
3. The presence of micro-organisms is often required to provide the enzymes.	No such requirement.
4. Are linked together in chains called metabolic pathways.	This type of inter-relation between inorganic reactions may also exist in certain production processes—see e.g. gasification—but it is not common.

All living organisms are divided into **two Empires**, each of which consists of **Kingdoms**. The two Empires are **Empire Prokaryota** and **Eukaryota**. The first Empire consists of the two Kingdoms of **Bacteria and Archaea**, which are made up of unicellular organisms only. Empire Eukaryota consists of the Kingdoms Protozoa, Chromista, Plantae, Fungi and Animalia, which may be either unicellular or multicellular, and includes the plants and animals.

The cells of organisms are basically similar. The cell structure of the basic eukaryotic cell consists of

(i) the cytoplasm, a fluid medium contained within a cell membrane (also called plasma membrane) all of which is contained within the cell walls in the case of the plant cell;
(ii) several organelles contained within the cytoplasm, including the cell nucleus, which contains the DNA molecules, and the mitochondria, which produce ATP. Each of these organelles is bound within its own membrane.

In eukaryotic cells, some of the energy production (i.e. ATP synthesis) occurs within the cytoplasm. However, most of it takes place within the mitochondria, which are (membrane-bound) organelles present in the cytoplasm.

The unicellular archaea (e.g. the **methanogens**) do not have membranes, and thus no organelles can form within their cytoplasm. Thus no mitochondria, cell nuclei, or other organelles exist within the archaea cells. The cytoplasm contains all the molecules normally contained within the organelles in one big common pool.

The ATP synthesis that normally proceeds in the mitochondria of eukaryotic cells cannot take place in the cells of the archaea as they do not have the mechanism, provided by the mitochondrial structure, for oxygen (even if it were present) to accept electrons.

The archaea are also obligate anaerobes. This means they can only survive in environments that lack oxygen. These organisms therefore must have some other form of electron acceptors. These can be molecules or ions.

4.3.1.4 What is cellular metabolism?

Biological organisms need to maintain and reproduce the cells that provide their basic structural units, as well as interacting with their internal and external environments. To fulfil these functions, they require energy, as well as the organic starting materials for the synthesis of the complex molecules (e.g. macromolecules) that constitute the building blocks of the cellular structure.

Biochemical reactions must cater for such requirements. These reactions collectively form the **cellular metabolism**, and comprise all the metabolic pathways required for

- the production of the energy needs of the organism, and
- the synthesis of the complex molecules for growth and reproduction from organic and inorganic precursors.

The metabolic pathways are conveniently grouped into two groups: **catabolic pathways**, which break down the macromolecules present within the organism to

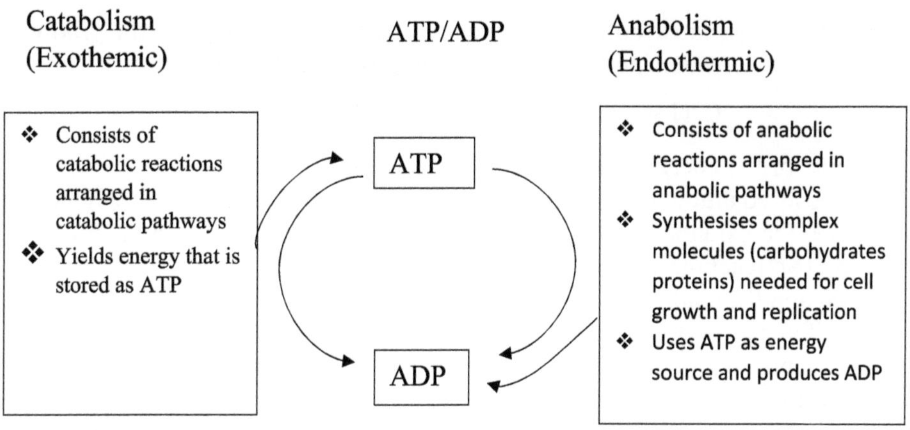

Figure 4.4. A schematic view of the two components of cellular metabolism. The exothermic catabolic pathways break down complex molecules and produce ATP. The endothermic anabolic pathways use the energy in ATP and organic and inorganic substrates to synthesise complex molecules.

produce the energy requirements, and **anabolic pathways**, which use simple organic molecules and inorganic nutrients from the environment as starting points (substrates) to build the complex macromolecules required by the organism.

Catabolic reactions are exothermic, and the energy produced is stored in **adenosine triphosphate (ATP)**, an energy-rich molecule. Anabolic reactions are endothermic, and use the energy stored in ATP to synthesise the complex molecules required for the organism's cellular structure and reproduction. Figure 4.4 demonstrates this division graphically.

The catabolic pathways describe the chains of reaction involved in the process of **respiration**.

4.3.2 Respiration and ATP

4.3.2.1 What is respiration?

Respiration is the process by which biological organisms acquire energy for their metabolic needs. The process consists of several metabolic pathways linked together, and the energy that is acquired is stored in ATP, the main end product of respiration. ATP is available as a source of energy for many metabolic processes within the organism, and is sometimes viewed as the 'energy currency' for metabolic processes.

The energy for ATP synthesis is obtained by the catabolism of glucose or other substrates such as amino acids, fatty acids and glycerol. These substrates for respiration are usually derived from the breakdown of complex polymers that include the polysaccharides, lipids, proteins and the nucleic acids (DNA and RNA). Most of the energy produced is used to synthesise ATP molecules from ADP molecules and (inorganic) phosphate ions, while the remainder is lost to the environment in keeping with the second law of thermodynamics.

It is important to stress that respiration is essentially an exothermic reaction, similar (in principle) to the combustion reactions discussed in chapter 3. When the substrate is glucose (a carbohydrate), the final result of respiration is the production of carbon dioxide, water and energy according to the equation

$$C_6H_{12}C_6 + 6C_2 \rightarrow 6CO_2 + 6H_2C + \Delta E \tag{4.4}$$

where most of the energy (ΔE) produced is used in ATP production.

The above equation is similar to that describing the combustion of glucose in oxygen (i.e. the oxidation of glucose), with the heat of combustion ΔH_c replaced by ΔE. These two reactions may be viewed as redox processes, where the fuel or substrate is oxidised and the oxidising agent (oxygen or some other molecule or ion) is reduced.

The difference between direct combustion and respiration is that, unlike the combustion process, the carbon dioxide and water are produced at different stages of the respiration process, and most of the energy produced in the respiration is stored as chemical energy in ATP (and not released to the environment as heat energy as in the combustion reaction). The other major difference of course is the involvement of enzymes in respiration.

These differences in the chemical and biochemical oxidation of glucose are important to note. Unlike the chemical oxidation where the reaction is a direct process, respiration requires specific mechanisms for the transfer of electrons and ions between the reacting species. This involves the participation of special electron carrier molecules.

4.3.2.2 Stages of aerobic respiration

Individual metabolic pathways make up distinct stages of respiration, and the number of ATP molecules produced per molecule of the substrate (the starting energy-rich molecule for the process) is not the same for all stages. This amount also depends on whether oxygen is available, i.e. whether it is **aerobic** or **anaerobic** respiration.

The process of **fermentation** occurs when oxygen is unavailable to the organism. The ATP yield can be considerably lower in this case. To obtain a holistic picture of respiration, it is best to begin by considering the **aerobic respiration of glucose**.

Aerobic respiration can be conveniently divided into three parts:

- Glycolysis;
- The Krebs Cycle (or the Citric Acid Cycle, sometimes also called the TCA cycle);
- Oxidative phosphorylation.

Each of these consists of chains of reactions involving the oxidation/reduction of electron-transporting molecules (called co-enzymes or co-factors) and the production of ATP. The molecules involved in the electron transport are:

- Nicotinamide Adenine Dinucleotide (NAD);
- Flavin Adenine Dinucleotide (FAD);
- and their reduced forms (NADH and $FADH_2$).

4.3.2.3 Glycolysis

Glycolysis is the breakdown of the 6-carbon glucose molecule into two 3-carbon pyruvate anions (or two pyruvic acid molecules) and the net production of two ATP molecules.

(See exercise 4.2—Respiration.)

The process takes place in the cell cytoplasm, and involves 10 steps, each catalysed by its own enzyme. It is divided into two phases. The first (called the **preparatory phase**) uses up two molecules of ATP to convert a glucose molecule to two glyceraldehyde-3-phosphate (GADP or GAP) molecules.

$$\text{Glucose} + 2\text{ATP} \rightarrow 2\text{GADP} \tag{4.5}$$

In the second (**pay-off phase**), the two GADP molecules are converted to two pyruvate and four ATP molecules.

$$2\text{GADP} \rightarrow 2\text{pyruvate} + 4\text{ATP} \tag{4.6}$$

There are thus a net of two ATP molecules produced, together with other co-enzymes that help in the transport of electrons. These include two NADH molecules and two H^+ ions.

The pyruvate then enters the mitochondria, where it reacts with Co-enzyme A (CoA) to form Acetyl-CoA.

$$\text{Pyruvate} + \text{CoA} \rightarrow \text{Acetyl–CoA} \tag{4.7}$$

This is facilitated by NAD^+, which forms two NADH molecules in the process, (accompanied again by two H^+ ions). Two CO_2 molecules are also produced in the process.

4.3.2.4 The citric acid cycle

Acetyl-CoA now participates in the citric acid cycle (which occurs inside the mitochondrion) to produce six NADH, two $FADH_2$, two ATP and four CO_2 molecules in total, together with six H^+ ions.

(See exercise 4.3—Aerobic respiration.)

4.3.2.5 Electron transport chain and oxidative phosphorylation

The NADH and $FADH_2$ that were produced in the preceding pathways are (energetic) electron carrier molecules. As mentioned earlier, respiration can be viewed as a redox process. This entails the transfer of electrons from the oxidising agent to the reducing partner in the reaction. The two molecules above serve the function of delivering the electrons from the substrate to the main site where the other partner in the reaction is present. In the case of aerobic respiration, this is oxygen, which acts as the **terminal electron acceptor** for the respiratory process.

The main aim of respiration is to produce ATP, and the bulk of this task is performed across the mitochondrial membrane in the process known as **oxidative phosphorylation**. For this process to proceed, a total of ten NADH, two $FADH_2$ molecules and ten H^+ ions are transferred to the reaction site in the mitochondrion where

- oxygen accepts the electrons and reacts with the hydrogen ions to form water;
- ADP reacts with an inorganic phosphate (P_i) to produce ATP according to the reaction.

$$3\text{ADP} + 3\text{Pi} + \Delta E \rightarrow 3\text{ATP} \tag{4.8}$$

4.3.2.6 Summary of aerobic respiration

The process of aerobic respiration may be summarised as follows:

- The main aim of aerobic respiration is to produce energy-rich ATP molecules that store energy for use in various metabolic processes in the cell.
- The metabolic pathway for aerobic respiration consists of the three sub-pathways of glycolysis, citric acid cycle and oxidative phosphorylation.
- There is a need to transfer energetic electrons from the substrate at the beginning of the process and to receive them at the end of the process. Ten NADH and two $FADH_2$ molecules are transferred to the mitochondria for the first purpose. The latter function is fulfilled by oxygen.

The production of ATP occurs as follows:

- two during glycolysis;
- two in the citric acid cycle; and
- thirty-four during oxidative phosphorylation at the end of the metabolic chain.

This amounts to the production of a total of 38 ATP molecules per glucose molecule. Figure 4.5 summarises the process diagrammatically.

(See exercise 4.4—Respiration and oxidation of glucose.)

4.3.3 Anaerobic respiration and fermentation

In the aerobic respiration of glucose, the first step (i.e. the glycolysis of the substrate into two pyruvic acid molecules) is dependent on the presence of the NAD^+ co-enzyme. This converts to the reduced NADH state, which provides a (high energy) electron carrier for the final stage of the whole metabolic process. For glycolysis to be sustained, a continuous supply of NAD^+ should therefore be available.

Under the normal conditions of aerobic respiration, this supply is obtained from the re-conversion of NADH to NAD^+ in the oxidative phosphorylation stage of respiration, after oxygen accepts electrons at the end of the electron transport chain. When oxygen in not available, this supply of NAD^+ for the glycolysis process is cut off. Unless an alternative supply is found, the whole respiration process will come to a stand-still.

Such alternative electron accepters may be available in the environment, or consist of molecules that are synthesised through a fermentation process. Examples of the former are inorganic ions such as NO_3^-, SO_4^{2-} or CO_3^{2-}. These ions may be utilised by micro-organisms that have the appropriate enzyme.

Examples of organic molecules that act as electron acceptors are ethanol and lactic acid. These molecules are produced by the processes of **alcohol fermentation** and **lactic acid fermentation**, respectively [10]. In lactic acid fermentation, pyruvate is converted to lactic acid which acts as the electron acceptor. This type of fermentation can be carried out by **lactobacillus bacteria**, which have the required enzyme to carry out the fermentation.

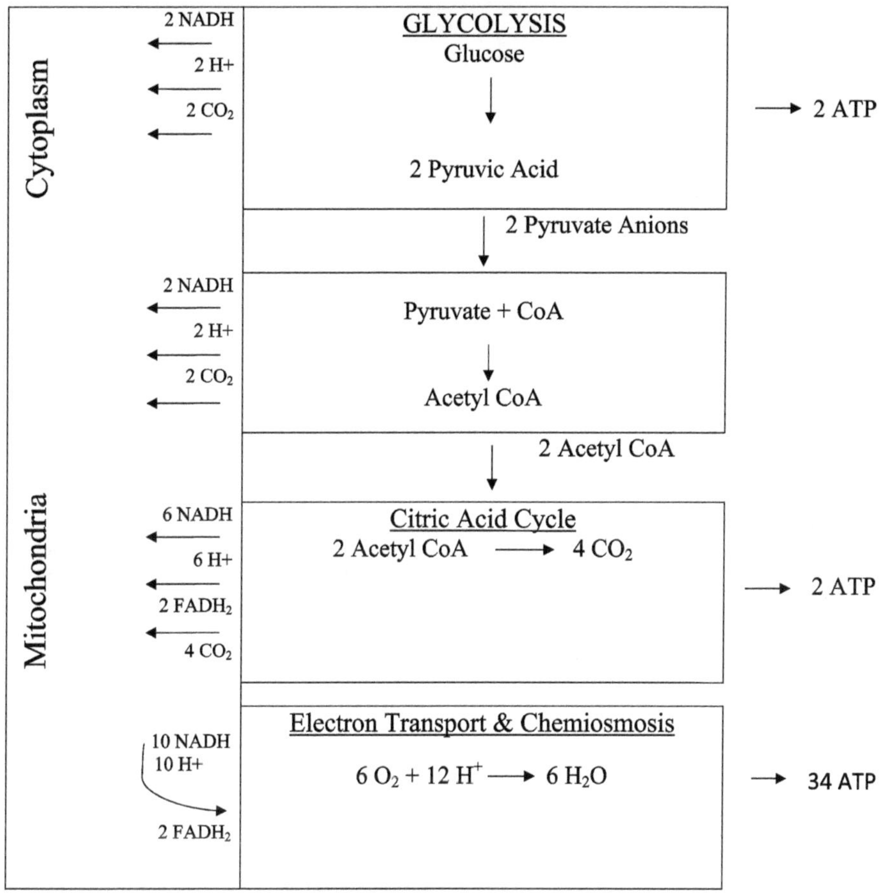

Figure 4.5. Stages of aerobic respiration.

In alcohol fermentation, the pyruvate is converted to ethanol, and the process is carried out by a strain of yeast bacterium known as **saccharomyces cerevisiae**. In all these cases where oxygen is not available, the new electron acceptor receives electrons from NADH, which converts to NAD^+ in the process and allows glycolysis to proceed.

Note that in the absence of oxygen, both the citric acid cycle and the oxidative phosphorylation pathways are excluded from the chain of metabolic pathways for respiration. The consequence is the reduction in the number of ATP molecules produced per glucose molecule. The exact numbers of ATP that will be produced will depend on the nature of the electron acceptor, but will always be lower than 38. It can indeed be as low as two ATP molecules.

Figure 4.6 provides typical examples of possible anaerobic pathways and the associated bacteria.

Table 4.3 provides a summary of the ATP yields for the various types of respiration and fermentation processes considered above.

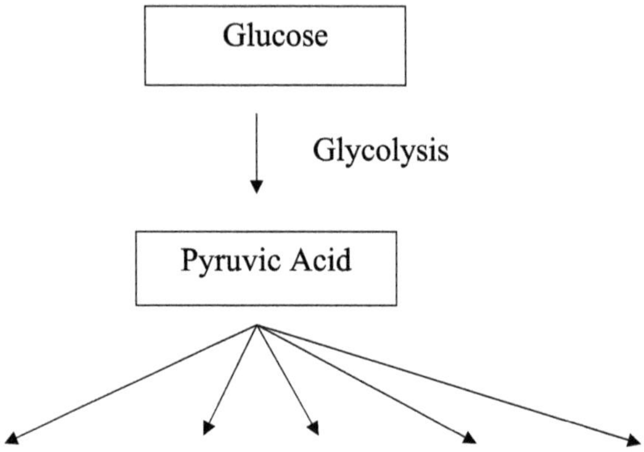

Bacterium	Saccharomyces (Yeast)	Streptococcus Lactobacillus	Clostridium	Escherichia, Salmonella	Enterobacter
Fermentation Product	Ethanol, CO_2	Lactic Acid	Butyric acid, Butanol, Acetone, Isopropyl alcohol, CO_2	Ethanol, Lactic Acid, Succunic acid, Acetic acid, CO_2, H_2	Ethanol, Lactic acid, Formic acid, Butanediol, Acetoin, CO_2, H_2

Figure 4.6. Some fermentation pathways and associated bacteria. Adapted from [10]. Copyright © 2001 John Wiley & Sons, Ltd. All rights reserved.

Table 4.3. Summary of ATP molecules produced from aerobic, anaerobic respiration and fermentation.

Type of process	Terminal electron acceptor	Number of ATP molecules produced
Aerobic respiration	Oxygen molecules	38
Anaerobic respiration	NO_3^-, CO_3^{2-}, SO_4^{2-} and other anions	More than 2 but fewer than 38
Fermentation	Ethanol or lactic acid molecules	2

4.3.3.1 Summary of aerobic and anaerobic respiration

Aerobic respiration is the production of ATP in the presence of oxygen, and includes the sub-processes of glycolysis, citric acid cycle and oxidative phosphorylation. In the case where the substrate is glucose, the process begins with the splitting of the glucose molecule to produce two pyruvic acid molecules (also called pyruvate), and ends with oxidative phosphorylation, where oxygen accepts electrons from electron carriers at the end of the electron transport chain. The whole process produces 38 ATP molecules per glucose molecule.

Anaerobic respiration takes place in the absence of oxygen. The metabolic pathway does not contain the citric acid cycle or the electron transport chain, and the number of ATP molecules produced varies between two and a number smaller than 38. The final electron acceptors in anaerobic respiration can be inorganic anions, or molecules, such as ethanol or lactic acid, that arise through the process of fermentation.

4.4 Anaerobic digestion and methane fermentation

4.4.1 Introduction

We have seen how ethanol, an alternative energy liquid biofuel, can be produced as a by-product of anaerobic respiration. Another green fuel product of anaerobic respiration is gaseous methane. This occurs as a component of biogas, which is a mixture of methane and carbon dioxide. Biogas is used for cooking and heating at the domestic level, and transportation and power generation within the commercial and industrial sectors of a country.

In contrast to alcoholic fermentation where a very limited number of micro-organisms are involved, this gaseous fuel is produced by the action of consortia of microbes on solid biomass. The first group of such micro-organisms break down (i.e. hydrolyse) biopolymers such as carbohydrates, lipids and proteins into their short-chain precursors (e.g. monomers such as sugars, fatty acids and amino acids). This is followed by conversion of these soluble compounds to volatile fatty acids and other compounds by a second cohort of bacteria. These compounds undergo further processing to produce acetic acid (which yields the acetate anion). In the last step, methanogenic bacteria convert the acetate into methane through the process of methane fermentation.

All these steps are biochemical processes facilitated by specific consortia of bacteria that provide the necessary enzymes for the reactions. The whole process is variously referred to as anaerobic digestion or methane fermentation. The individual steps in the digestion of biomass to produce methane may be summarised as follows [11].

4.4.1.1 Stage 1—hydrolysis

Large, insoluble biopolymers consisting of carbohydrates, lipids and proteins are broken down (depolymerised, or hydrolysed) into their (soluble) monomers by hydrolytic bacteria that possess the hydrolytic enzymes known as cellulases, hemi-cellulases, amylases, lipases and proteases. The products include monosaccharides (sugars) arising from the polysaccharides, fatty acids and glycerol resulting from the cracking of the fatty acid chains from their glycerol backbone in triglycerides, and amino acids produced from the depolymerisation of proteins.

4.4.1.2 Stage 2—acidogenesis

Acidogenic bacteria convert the above products into volatile (i.e. low molecular weight) fatty acids including acetic, propionic, butyric and lactic acids, alcohols (methanol and ethanol), as well as hydrogen and carbon dioxide.

4.4.1.3 Stage 3—acetogenesis

The molecular products of acidogenesis are metabolised by acetogenic bacteria to produce acetic acid (also called acetate), hydrogen and carbon dioxide.

4.4.1.4 Stage 4—methanogenesis

In the last step, methanogenic bacteria convert the acids, carbon dioxide and hydrogen into methane and carbon dioxide.

4.4.2 Methane fermentation pathways

Although still referred to as a fermentative process, methanenogenesis is now considered to differ from both fermentation and respiration [12].

There are three metabolic pathways associated with methanogenesis [11–13]. They are the:

- aceticlastic/acetotrophic pathway;
- hydrogenotrophic pathway;
- methylotropic pathway.

In the first pathway, the methanogenic bacteria convert the acetic acid substrate into carbon dioxide and methane. In the hydrogenotrophic pathway, carbon dioxide produced in stage two is reduced to methane, while in the third pathway, methanol and methylamine are reduced.

These metabolic pathways for methane fermentation are still under active investigation. Further details and references can be found in the work of Sikora *et al* (2018) [11] and Zhu *et al* (2020) [13], who have prepared lists of enzymes responsible for the pathways, and applied genome-centric metagenomics and metatranscriptomics to their study.

Figure 4.7 provides a summary of the biochemical processes of respiration and fermentation processes considered in this section.

4.5 Electrochemical conversion of biofuels

4.5.1 Introduction—thermal vs electrochemical conversion of chemical energy

The previous sections of this chapter considered the conversion of bioenergy into forms that were more amenable to use by the end user. Section 4.2 considered the production of biodiesel, HVO and synthetic fuels from biomass through chemical means, while the following two sections introduced the biochemical processes for the production of bioethanol and biogas.

The end product of all these processes was the conversion of biomass into new fuels, known as **secondary fuels**. As is well known, however, one of the most important forms of energy used by the consumer is electrical energy. An energy conversion technique that could convert bioenergy directly into electricity is therefore of significant interest.

Electrochemical conversion of energy is the conversion of chemical energy directly into electricity. The energy content of a biofuel is contained in the chemical bonds of its molecules. One means of converting this energy into electricity is via the use of **thermal power generation**. In such technology, the chemical energy of the biofuel is first converted into heat energy through combustion. This heat energy is then fed into a heat engine which drives an electrical generator.

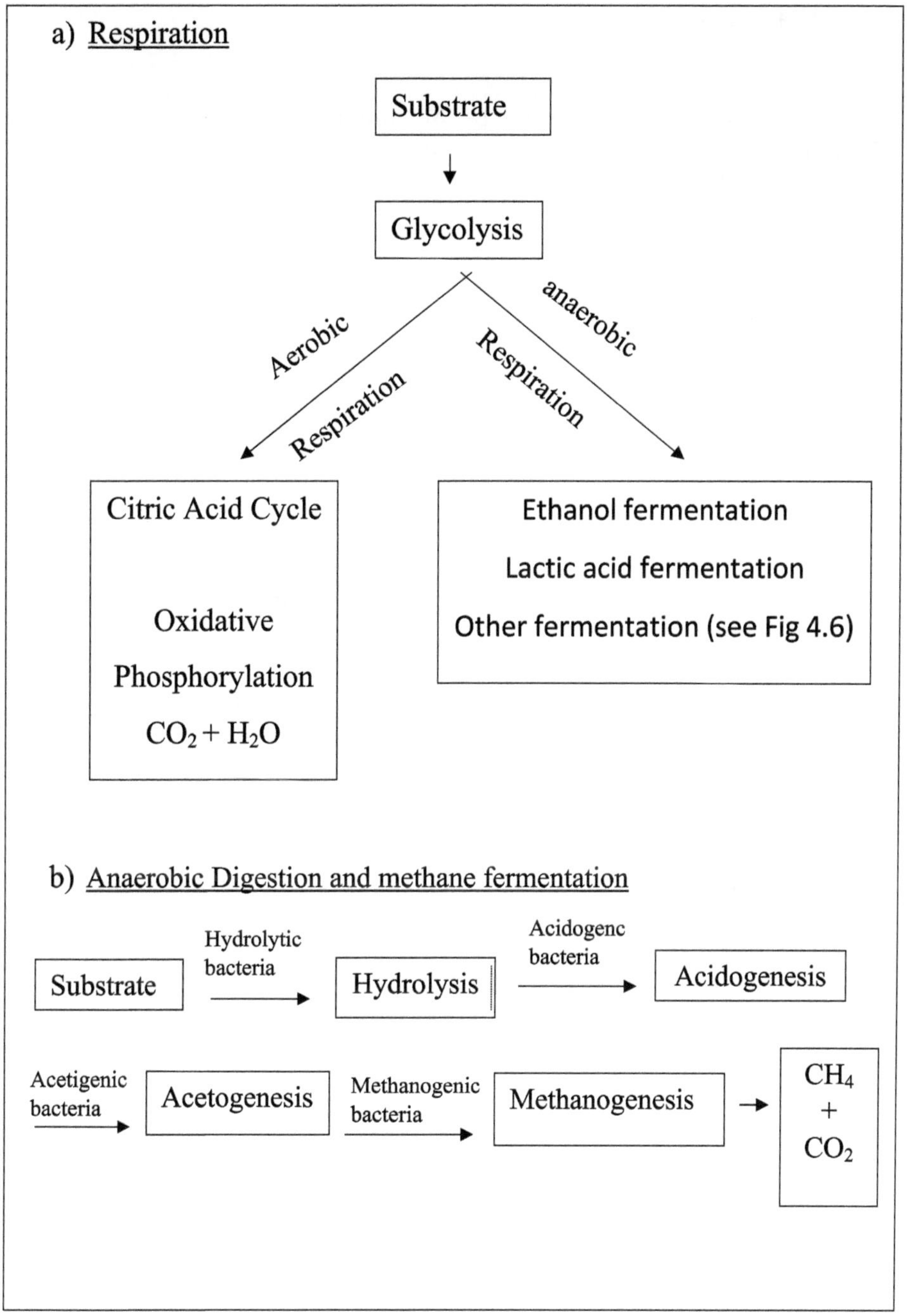

Figure 4.7. A holistic overview of the respiration and fermentation processes producing alcohol and methane as end products: (a) aerobic and anaerobic respiration; (b) anaerobic digestion and methane fermentation.

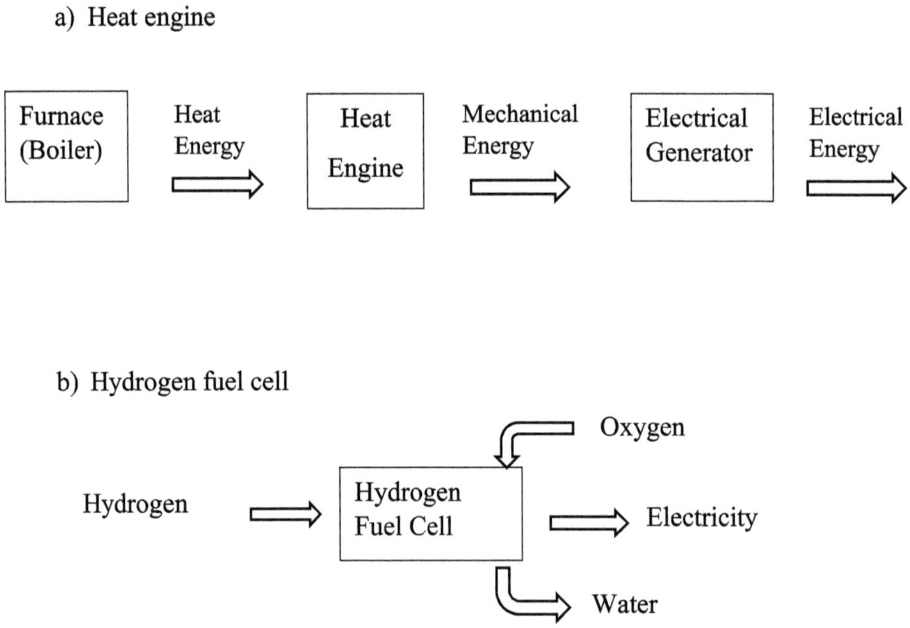

Figure 4.8. Producing electricity from the chemical energy stored in a biofuel through the use of (a) a heat engine and (b) a fuel cell. In the heat engine, the biofuel is burned (i.e. oxidised) to produce heat energy, which in turn is fed into a heat engine. The engine converts the heat energy to work (which is later converted to electricity) according to the second law of thermodynamics. In the hydrogen fuel cell, the chemical energy is converted directly into electricity.

However, in the case where the fuel is hydrogen, the chemical energy can be converted directly into electricity using a device known as a **hydrogen fuel cell** [14]. The difference between the two ways of converting chemical energy into electricity is demonstrated graphically in figure 4.8.

The nature and applications of fuel cells will be discussed at greater length in chapter 7.

4.5.2 The importance of fuel cells

There are two reasons why fuel cells may be preferred over heat engines for electricity generation. Firstly, because they avoid the use of heat energy as the source of input energy for the electricity generation device, their efficiencies can be much higher than those of heat engines. (This will become clearer in chapter 5 where the topic of thermodynamics and heat engines is discussed at length.)

In addition to their higher efficiencies as compared to thermal power generation, fuel cells are also important because of their potential for **climate change mitigation**. Because they do not require heat energy as an intermediate step (as in the production of electricity by thermal power plants), such devices do not produce **greenhouse gases** such as carbon dioxide. They are thus climate-friendly and provide good candidates as green technology for electricity production.

4.5.3 Blue hydrogen and hydrogen fuel cells

It was briefly mentioned in chapter 3 that blue hydrogen, produced from coal-derived syngas, could be used as the energy source of fuel cells. This form of hydrogen is derived from the gasification of coal and natural gas (the latter of which consists largely of methane gas). Other sources of hydrogen include the gasification of biomass and biochemical processes such as anaerobic digestion.

In a hydrogen fuel cell, hydrogen is reacted with oxygen (available in air) with the production of electricity and the release of water as the only products according to the equation

$$2H_2 + O_2 \rightarrow 2H_2O + \Delta E \tag{4.9}$$

where most of ΔE appears as electrical energy.

However, unlike normal oxidation reactions, the two reactants are not mixed together physically, but are kept apart at two different electrodes. Charge transport between the reactants is facilitated via an electrolyte separating the two electrodes and an external (electron-conducting) circuit.

It must be noted that unlike hydrogen derived from 'green' sources, coal or fossil fuel-derived hydrogen (called blue hydrogen), generally produces carbon dioxide as well. Thus, unless steps have been taken to satisfactorily dispose of the carbon dioxide produced, fuel cells using blue hydrogen generally do not qualify as green technology. Some of the methods for the disposal of the carbon dioxide currently being proposed are carbon capture and storage in geological basins and oil fields [15, 16]. (The term 'grey hydrogen' is sometimes used to distinguish fossil fuel-derived hydrogen that still contains carbon dioxide from hydrogen that does not. See chapter 11 for further elaboration.)

(See exercise 4.5—Efficiency of electrochemical energy conversion.)

The process of electrical energy production in a hydrogen fuel cell may be viewed as the reverse of the electrolysis of water. In the latter process, electrical energy is provided via two electrodes (anode and cathode) placed in an electrolyte, which results in the production of hydrogen and oxygen gases at the electrodes. In a hydrogen fuel cell, hydrogen and oxygen are introduced at the electrodes of the fuel cell. These are converted into water at one of the electrodes with the production of electricity in the external circuit.

Many types of fuel cells are in existence. They have a long history of application, dating back beyond the Apollo missions to the moon where they were used for the on-board power requirements of the spacecraft. The technology of fuel cells will be discussed further in part II of the book.

Exercises and questions

Exercises

Exercise 4.1. Calculating biodiesel yield in transesterification.

a) Inspect figure 4.2 and calculate the biodiesel yield as follows:

(i) Determine the number of moles of methanol required to react completely with one mole of a triglyceride, and the number of moles of biodiesel and glycerol produced.

(ii) Assume that the triglyceride is triolein, where all three fatty acid chains R_1, R_2, R_3 derive from oleic acid, which has 18 carbon atoms and one double bond in its R group. Calculate the masses of the feedstock and methanol required to produce one kilogram of biodiesel. Assume that the atomic masses of carbon, hydrogen and oxygen are 12.0, 1.0 and 16.0, respectively.

b) (i) In the practical production of biodiesel, an amount of methanol exceeding the amount calculated above will be used. Why is this necessary?

(ii) List the substances you expect to find in the reaction vessel once the reaction is over, and state why they are present.

Exercise 4.2. Respiration.

ATP is a molecule consisting of an adenine molecule and a triphosphate chain linked to separate carbon atoms of a ribose molecule.

a) Draw the structure of the ATP molecule, clearly labelling its constituent parts.

b) The last phosphate anion on the triphosphate is loosely bound and can be easily removed. In the first step of glycolysis, an ATP molecule phosphorylates the glucose ring by transferring its last phosphate group to the hydroxyl group on carbon 6 of glucose to produce glucose-6-phosphate. Show an equation describing this process and name the enzyme mediating the process.

c) Describe, with the help of an equation, how ATP is synthesised in the mitochondria of eukaryotes from ADP. What is the process called, and what is the name the enzyme that mediates the process?

Exercise 4.3. Aerobic respiration.

Identify the function of each of the molecules (NADH, FADH2 and ATP) produced during aerobic respiration. Indicate how the first two molecules are produced.

Exercise 4.4. Respiration and oxidation of glucose.

Show how figure 4.5 accounts for all the molecules in the overall oxidation of glucose given by equation (4.4) ($C_6H_{12}O_6 + 6O_2 \rightarrow 6CO_2 + 6H_2O + \Delta E$).

Exercise 4.5. Efficiency of electrochemical energy conversion.

In a fuel cell, chemical energy is converted directly into electricity. Explain why, in a practical setup, not all of the energy would appear as electricity. (*Hint*: think of a practical application of a fuel cell to generate electricity, and consider the energy losses due to the inefficiencies.)

Questions

Question 4.1. Properties of diesel engine fuels.

Identify at least four important properties an organic liquid should have to qualify as a diesel engine fuel. Give examples of issues that occur when a fuel does not meet these fuel criteria.

References

[1] Lipid biochemistry. Britannica [accessed 7 November 2022]. Available from https://britannica.com/science/lipid

[2] Historical perspectives on vegetable oil-base diesel fuels. AOCs Lipid Library. Available from https://lipidlibrary.aocs.org/resource-material/scientific-milestones-in-lipid-research/historical-perspectives-on-vegetable-oil-based-diesel-fuels

[3] Van Gerpen J, Shanks B, Pruszko R, Clements D and Knothe G 2004 *Biodiesel Production Technology* (Golden, CO: NREL), 110 Report No. NREL/SR-510-36244. Available from https://nrel.gov/docs/fy04osti/36244.pdf

[4] Singh A and Deo R (ed) 2020 *Translating the Paris Agreement into Action in the Pacific.* (Cham: Springer Nature) p 323

[5] Neste Worldwide 2016 *Neste Renewable Diesel Handbook* (Rotterdam: Neste Propriety Publication) p 57

[6] Pearlson M N 2011 A techno-economic and environmental assessment of hydroprocessed renewable diesel fuels *MSc Thesis* (Cambridge, MA: M.I.T.) p 106

[7] Van de Loosedrecht J, Botes F G, Ciobica I M, Ferreira A, Gibson P and Moodley D J *et al* 2013 Fischer–Tropsch synthesis: catalysis and chemistry. J Reeddijk and K Poepellmeier *Comprehensive Inorganic Chemistry II* vol 7 (Oxford: Elsevier) ch 7.20 pp 525–57

[8] Biosphere. Environmental Conditions. Britannica [accessed 20 February 2022]. Available from https://britannica.com/science/biosphere/Environmental-conditions

[9] Ruggiero M A, Gordon D P, Orrell T M, Bailly N, Bourgoin T, Brusca R C and Cavalier-Smith T *et al* 2015 A higher level classification of all living organisms *PLoS One* **10** e0119248

[10] Müller V 2001 *Bacterial Fermentation* (New York: Wiley) Available from https://web.archive.org/web/20140908223508/http://web.oranim.ac.il/courses/microbiology/Bacterial%20Fermentation%20Nature.pdf

[11] Sikora A, Detman D, Meilecki D, Chojnacka A and Blaszczyk M 2018 *Searching for Metabolic Pathways of Anaerobic Digestion: A Useful List of the Key Enzymes* (London: IntechOpen) [accessed 20 February 2022]. Available from https://intechopen.com/chapters/63744

[12] Fenchel T, King G M and Blackburn T H 2012 *Bacterial Biogeochemistry* 3rd edn (Amsterdam: Elsevier)

[13] Zhu X, Campanaro S, Treu L, Seshadri R, Ivanova N and Kougias P G *et al* 2020 Metabolic dependencies govern microbial syntrophies during methanogenesis in anaerobic digestion ecosystem *Microbiome* **8** 1–14

[14] Sorensen B and Spazzafumo G 2018 *Hydrogen and Fuel Cells* 3rd edn (Amsterdam: Elsevier) p 522 see also https://mypages.iit.edu/~smart/garrear/fuelcells.htm

[15] IEA. About CCUS [accessed 20 February 2022]. Available from https://iea.org/reports/about-ccus

[16] Carbon capture, use and storage. UNFCCC [accessed 20 February 2022]. Available from https://unfccc.int/resource/climateaction2020/tep/thematic-areas/carbon-capture/index.html

Part II

The physics and technology of bioenergy use

IOP Publishing

Bioenergy for Power Generation, Transportation and Climate Change Mitigation

Anirudh Singh

Chapter 5

The thermodynamics of heat engines

A significant amount of bioenergy is used to generate mechanical and electrical power through the use of heat engines, and it is therefore important to understand how heat engines work. This chapter provides a comprehensive introduction to the principles of thermodynamics as it applies to the operation of heat engines, and considers the specific examples of heat engines based on the Rankine and Brayton cycles.

It begins by introducing the language of thermodynamics and shows how a heat engine uses cyclic processes to convert heat into work according to the first and second laws of thermodynamics. The heat engine power cycles are then introduced, followed by a detailed discussion of the Rankine cycle and its application to steam turbines. Realistic engines using super-heating and reheating are then investigated. The chapter ends with a brief look at the Brayton cycle and its application to gas turbines.

5.1 Introduction

We need to convert the chemical energy of biomass into other forms before it becomes available for use. These forms include heat energy for cooking and heating, mechanical energy for transportation and electrical energy which is arguably the most important form of energy in use today.

Burning of biomass converts its chemical energy store to heat energy. But a large share of a country's energy demand is for mechanical energy for transportation and the production of electricity. A quick assessment of the relative importance of mechanical and electrical forms of energy obtained from biomass can be obtained by considering the energy use scenario in the USA.

As table 2.3 shows, in 2020:

doi:10.1088/978-0-7503-3555-3ch5

- 28% of the country's biomass energy production was used in the transportation sector (to provide the mechanical power to drive vehicles);
- 9% by the power generation sector (i.e. pure electrical power);
- 50% by the industrial sector (as both electrical and mechanical power).

Assuming that the electrical and mechanical power needs of the industrial sector are equally divided, these statistics reveal that as much as 53% of the biomass energy utilised in the USA in 2020 was used towards the mechanical power needs of the country. Together with the 9% used for electrical power generation, this is strong indication that there is a high demand for the conversion of available biomass energy into mechanical and electrical forms on a global scale.

Two devices available today for the conversion of biomass energy into these more useful forms are the internal combustion engine and the steam or gas turbine. Both are heat engines, which convert the chemical energy of the biomass into mechanical energy through these devices according to the principles of thermodynamics.

This chapter considers the **basic theory of thermodynamics** as applied to the operation of heat engines. It establishes the thermodynamic principles on which the **heat engine** is based, and carries out a brief survey of the **types of heat engines** (including external and internal combustion engines, reciprocating engines and turbines) that are possible. The chapter ends with a look at the **Rankine cycle and its application to the steam turbine** that provides the main workhorse for the electrical power industry today.

The following chapter will consider how heat engine based power generation technologies are used in the generation of electrical energy from bioenergy sources.

5.2 Heat engine—basic concepts

A heat engine is a device that converts heat energy to mechanical energy in the form of work done on an external load. It performs this task by allowing heat to flow through it from a hot place (called a hot reservoir) to a cold place (called a cold reservoir) to which it is in thermal contact.

According to Fourier's Law of Thermal Conductivity [1], heat only flows between the two ends of a material medium if there is a finite temperature gradient between them. This is seen in the application of Fourier's Law (equation (5.1)) to two points in a material that are separated by a distance x:

$$Q = -kA[(T_1 - T_2)/x] \tag{5.1}$$

where Q is the thermal flux, A is the cross-sectional area of the ends separated by distance x and held at temperatures T_1 and T_2, respectively, and $\boldsymbol{k}$ is a constant called the **thermal conductivity**.

In this equation, the rate of heat flow Q is seen to be directly dependent on the term $(T_1 - T_2)/x$, called the **temperature gradient**. This demonstrates unambiguously that heat flow cannot occur between two places in a material medium unless there is a temperature difference between them.

Heat engines utilise this principle of heat transfer by providing a means of converting the heat flow between two thermal reservoirs into work done on the external environment. Figure 5.1 provides a schematic illustration of how a heat engine works in principle.

In this diagram, the source of the heat energy is the hot reservoir, which exists at the high temperature T_h. The amount of heat that is delivered by this reservoir to the heat engine during any fixed period of time Δt is Q_h.

The useful work performed by the engine during this time is denoted W_{eng}. Not all of the heat energy arriving from the hot reservoir is converted to work by the engine, and an amount Q_c is rejected to the heat sink, represented in the diagram as the cold reservoir, which exists at the low temperature T_c.

The principle of operation of heat engines can be understood in terms of the **internal energy** $\boldsymbol{U}$ **(or** $\boldsymbol{E_{int}}$**)** of a **system** that is taken through a **cyclic process** by the engine. During this cycle, the system is taken from an **initial state**, via processes

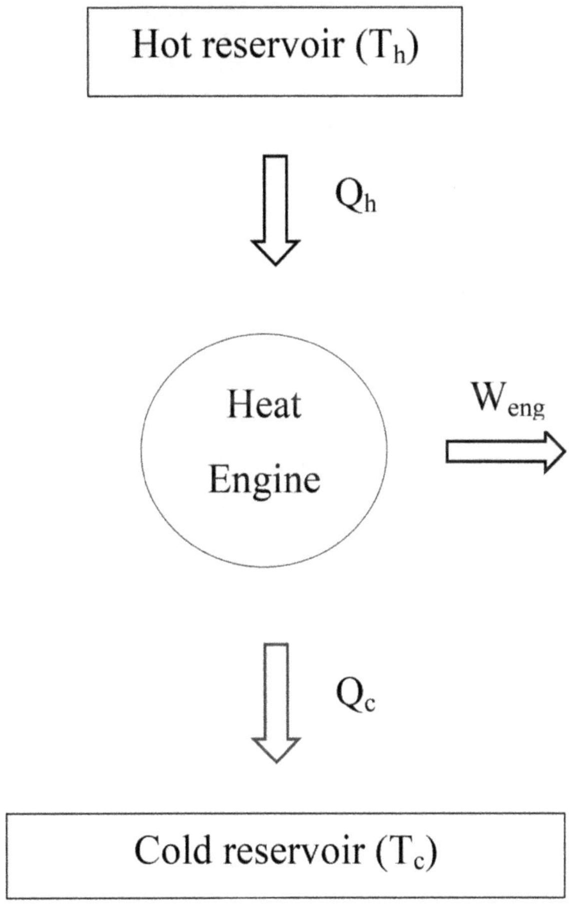

Figure 5.1. Schematic diagram illustrating the principle of converting heat energy into mechanical energy using a heat engine.

where it does work on the external environment (and thus changes its internal energy), and through further processes in which the state of the system (including its internal energy) is brought back to its initial state. All of these notions will be discussed more fully below.

There are several ways in which such a logical design of a heat engine can be implemented physically. The most common of these include the use of turbines and reciprocating (i.e. piston) engines. Figure 5.2 shows how turbines can convert heat energy to mechanical energy.

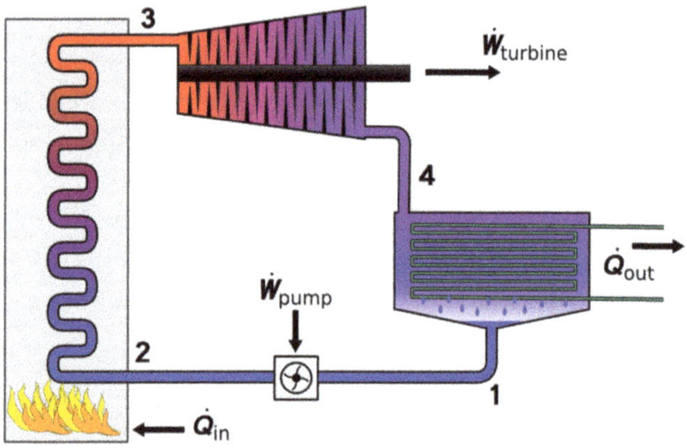

Figure 5.2. Schematic diagram of a steam turbine engine that converts heat energy into mechanical energy [2]. This Rankine cycle layout has been obtained by the author from the Wikimedia website where it was made available under a CC BY-SA 3.0 licence. It is included within this article on that basis. It is attributed to Andrew Ainsworth.

In this figure, the system is a small volume of the gas or liquid, called the **working fluid**, encased inside the sealed tubes of the engine as shown. This fluid is heated in the **boiler**, or **furnace** (the closed vessel showing the flame) and the hot pressurised steam produced is released into the **turbine** which turns as a result. The spent fluid is then condensed back to its original state in the blue condenser and pumped back to the boiler to complete the cycle. In operation, the heat engine repeats this cycle continuously until it is switched off.

5.3 Basic concepts of thermodynamics

We need to have knowledge of the basic terminology and principles of thermodynamics to be able to understand how heat engines really work. Specifically, we need to know what a system is, how it is described in terms of its state variables, and the physical processes that take a system from one state to another. We also need to realise that the law of conservation of energy must apply to thermodynamics, and is expressed as its first law. In addition to all this, we need to be aware that a second law also exists in thermodynamics which places theoretical upper limits to the maximum efficiency any heat engine can have.

5.3.1 System

In thermodynamics the **system** is any collection of objects and the energy related to it (i.e. any small part of the universe that is of interest to us). The rest of the universe is then treated as the **environment** of this system. In our case, we are interested in how a heat engine is able to perform work on its environment through the use of a **working fluid**. This fluid is thus chosen as the system, and the environment is the device where the electrical or mechanical energy appears (e.g. the electrical generator or the drive shaft).

5.3.2 State variables

The system must be described in terms of variables that can be measured using physical measurement techniques. These variables are **temperature (*T*), pressure (*P*), volume (*V*), internal energy (*U* or E_{int})**, and **entropy (*S*)**. Some or all of these will change as the system does work on the environment.

These variables describe the physical changes that are required to take the system from one state to another. The relation between such **processes** and the resulting changes in the system, taken to be a volume of gas in a cylinder, is shown in figure 5.3. In this figure, pressing the piston down the distance Δx increases the pressure inside the gas, and reduces its volume. Thus, the state of the system, which is described in terms of these state variables, changes as well.

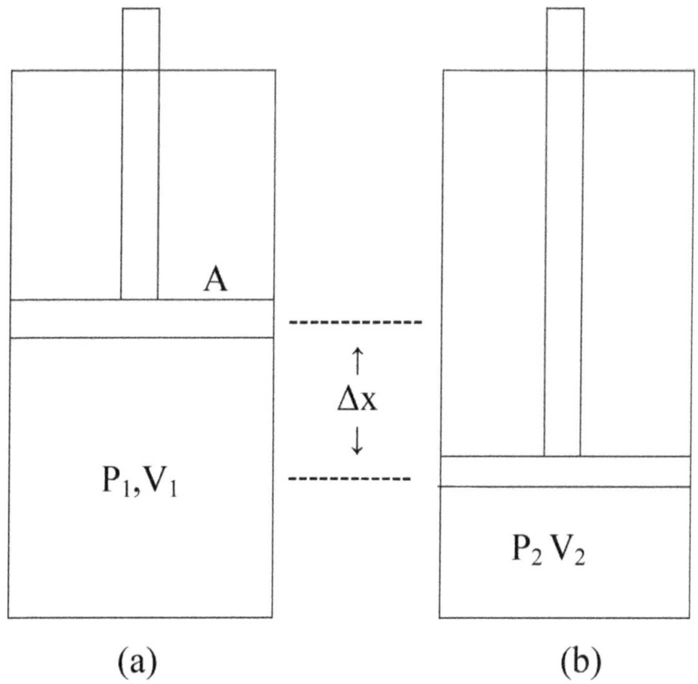

Figure 5.3. An example of a system (a gas contained in a cylinder) and a process that changes its state. The piston is used to apply pressure to the system which changes its volume.

5.3.3 State diagrams

The relationship between the states of a system and state variables, such as pressure and volume, can be displayed graphically in **state diagrams**. Figure 5.4 is a particular example of a state diagram, called a *P–V* diagram as it is expressed in terms of the state variables *P* and *V* [3].

Such diagrams are very useful in illustrating changes of states brought about by physical processes, as well as the assessment of the work done by the system.

5.3.4 Processes and paths

In state diagrams, each state of the system is represented by a point. A **path** is the line joining one state to another. It represents the locus of all states that the system goes through in going from the first state to the second.

As demonstrated in figure 5.3, changes in states are brought about by physical processes. Each physical process is represented by a unique path in the state diagram. Special instances of such processes are those that occur when certain state variables are held fixed. These special instances are summarised in table 5.1, and play important roles in the power cycles that determine the operation of heat engines.

(See exercise 5.1—States and paths.)

5.3.5 First law of thermodynamics

Energy must be conserved in any process. This is the same as saying that the change in the internal energy ΔE_{int} of a system must be equal to the difference between the energy added to the system and the energy it loses to the outside environment in any

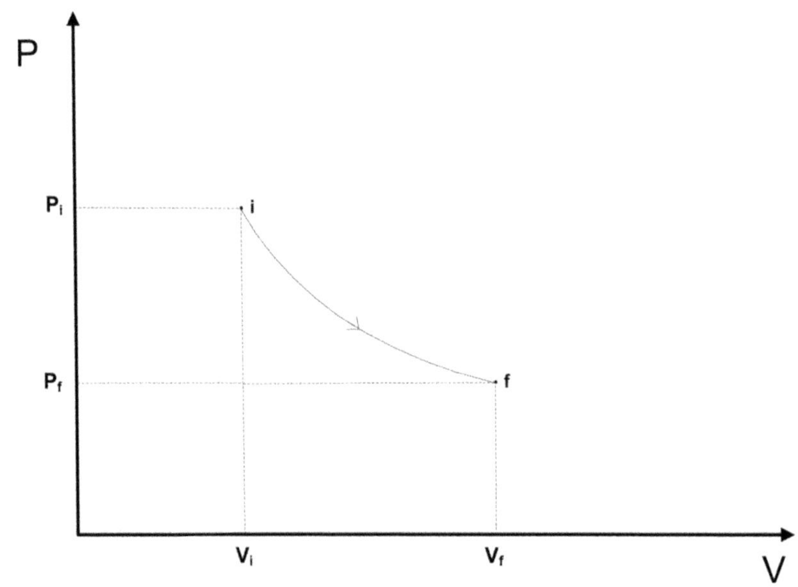

Figure 5.4. A *P–V* diagram (and example of a state diagram) showing the initial and final states of a system and the path leading from one state to the other.

Table 5.1. Special processes in thermodynamics.

Process	The state variable that is held fixed
Isobaric	Pressure P
Isothermal	Temperature T
Isochoric (isovolumetric)	Volume V
Isentropic (Adiabatic)	Entropy S

process. Now the internal energy of a system can be changed by adding heat Q to it or doing work W on it. If we choose a sign convention where the internal energy of a system increases when heat Q flows *into* the system, or when work W is done *on* the system, then the above statement may can be expressed as equation (5.2) [4].

(The First Law of Thermodynamics):

$$\Delta E_{\text{int}} = Q + W \tag{5.2}$$

where Q is positive when the heat is flowing ***into*** the system, and W is positive when work is done ***on*** the system. This is the **First Law of Thermodynamics** [4]. It is simply a re-statement of the law of conservation of energy in the context of changes in the internal energy of systems due to heat and work, using the mentioned sign convention.

5.3.6 Cyclic processes

So far, we have seen how the working fluid of a heat engine can be represented as a thermodynamic system which is described by its state variables. We have also seen that changes in the state of the system are brought about by physical processes, and that the first law of thermodynamics relates the change in the internal energy of the system to the heat flow in or out of the system and the work done to or by the system.

How these observations can be utilised in the conversion of heat to work is seen by considering a process that takes the system through several states, and finally brings it back to the original (starting) state. This is a situation where the final state is the same as the initial state. The combined result of these individual processes is called a **cyclic process**, and specific choices of the individual processes lead to the well-known cyclic processes. Figure 5.5 is an example of a cyclic process. If processes 1 and 3 are chosen to be isothermal and 2 and 4 are isentropic, this becomes the special cycle known as the **Carnot cycle**.

As the final state of the system is the same as the initial state in any cyclic process, no net change occurs to the internal energy of the system. Thus, for a cyclic process,

$$\Delta E_{\text{int}} = 0 \tag{5.3}$$

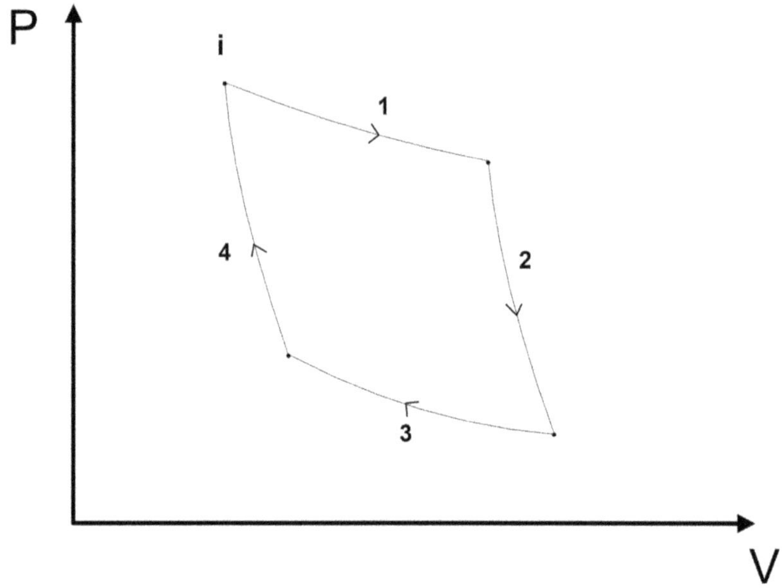

Figure 5.5. A cyclic process, in which a system is taken from an initial state *i*, via processes 1, 2, 3, 4, back to its initial state. In the Carnot cycle, processes, 1 and 3 are isothermal and 2 and 4 are isentropic.

Using the first law, this yields the important result that

$$Q = -W \tag{5.4}$$

This shows us that it is theoretically possible to use a cyclic process to convert heat energy Q into work W, and the amount of work done by the system is equal in magnitude to the heat added to it.

It can be shown that **the work done by a heat engine in a single cycle is equal to the area enclosed within the cyclic process on its *P–V* diagram**. As seen in figure 5.5, this area is enclosed within the paths representing the component processes of the entire cycle. As we can increase this area by choosing the processes judiciously, it is clear that different choices of cycles will represent different amounts of work done per cycle by the system.

(See exercise 5.2—Work done on a system.)

5.4 Principle of operation of the heat engine

5.4.1 New form of the first law

It was seen above that it is theoretically possible for a device to convert heat energy to work using a heat engine. To continue with the discussion, it is helpful to write the first law of thermodynamics in a different form.

In the current expression for the first law (equation (5.2)), W stands for the work done ***to*** a system, which is the working fluid of the heat engine in the present case. However, heat engines by their nature are devices that do work **on** the environment (e.g. an electric generator connected to the turbine of the heat engine). Thus, it is better to use an expression that describes this directly. This is done by defining the new quantity W_{eng} such that

$$W_{eng} = -W \tag{5.5}$$

Inspection of figure 5.2 shows that in a real heat engine, some work must be done on the working fluid to pump it back into the boiler. The work that is of interest, however, is the net amount of work done by the engine, which is found from the difference between the work output and work input for the system. This is considered by simply letting W_{eng} represent the net work done by the system.

The expression Q contained in the First Law of Thermodynamics for the heat transferred to the system also needs to be determined for the real heat engine. Figure 5.1 shows that the quantity Q_h is transferred to the system (the working fluid contained in the engine), and Q_c is rejected (or output) to the cold reservoir. Thus, the actual input of heat into the working fluid is the difference Q_{net} between these two quantities, i.e.

$$Q_{net} = Q_h - Q_c \tag{5.6}$$

This gives us the expression for the first law applicable to the real heat engine as:

The First Law of Thermodynamics for a real heat engine:

$$\Delta E_{int} = Q_{net} - W_{eng} \tag{5.7}$$

where the terms are as defined above. This new form of the first law has been used in the description of the heat engine in figure 5.1, and will be used in all discussions that follow.

5.4.2 The second law of thermodynamics

To fully understand the ability of heat engines to convert heat to work, one must realise that such devices must also obey the **Second Law of Thermodynamics**. There are various ways in which this law is stated, and the reader is encouraged to carry out a survey of all these forms. But for the current purposes, it is most appropriately stated as follows:

The Second Law of Thermodynamics:
It is impossible to convert all heat transferred to a heat engine into work.

Another statement of the second law that is relevant to the present purpose is that that **no heat engine can be 100% efficient**. Yet other forms involve the entropy of the system and the rest of the universe, but are less relevant to our purposes.

The second form of the law means that the heat Q_c rejected to the cold reservoir in figure 5.1 can never be equal to zero. Indeed, there is a theoretical maximum efficiency for any heat engine, called the **Carnot Efficiency**, that no physical heat engine can surpass. To see what this is, let us first see how the **thermal efficiency** of a heat engine is defined.

5.4.3 Thermal efficiency

The efficiency of a heat engine is defined, as usual, as the ratio of the output to the input energy of the system. The output is the work done W_{eng}, while the input is the heat input Q_h.

At the end of a complete cycle, the first law of thermodynamics gives

$$\Delta E_{int} = Q_{net} - W_{eng} = 0$$

Thus,

$W_{eng} = Q_{net}$. As the efficiency, by definition, is

$\eta_{Th} = W_{eng}/Q_h$. Its expression becomes

$\eta_{Th} = W_{eng}/Q_h = Q_{net}/Q_h = (Q_h - Q_c)/Q_h$. The final expression for the thermal efficiency therefore is:

Thermal efficiency of a real heat engine:

$$\eta_{Th} = 1 - Q_c/Q_h \tag{5.8}$$

(See exercise 5.3—Thermal efficiency of a heat engine.)

5.4.4 Carnot efficiency

The Carnot efficiency of a heat engine is the maximum possible efficiency attainable by a heat engine. This is worked out by considering a (hypothetically ideal) heat engine that uses the Carnot cycle, and is given by:

The Carnot Efficiency of a heat engine:

$$\eta_c = 1 - T_c/T_h \tag{5.9}$$

where T_c and T_h are the temperatures (measured in Kelvins) of the cold and hot reservoirs, respectively.

Thus, in addition to obeying the Second Law of Thermodynamics (which does not provide any specific upper limit to their efficiency) all real heat engines must also abide by the more stringent law of efficiency laid down by Carnot. This efficiency is easily calculated, and the efficiency of real heat engines is always lower than this limiting value.

5.5 Heat engine power cycles

As noted earlier, power cycles of heat engines are made up of processes, and the choice of the processes determines the type of power cycle. Power cycles are conveniently represented by state diagrams that show the specific paths taking the system from one state to another in the full cyclic process [5]. Each power cycle thus has its own characteristic state diagram.

(See exercise 5.4—Deducing paths of processes.)

To consider power cycles further, we can start with that of the ideal heat engine, which delivers the theoretical maximum achievable efficiency of any cycle. This is the **Carnot cycle**, and we have already discussed its representation in figure 5.5. The Carnot Cycle is made up of paths representing four component processes. Two of these are isothermal processes, while the remaining two are isentropic (i.e. adiabatic) processes.

This cycle gives the highest theoretical efficiency of any heat engine cycle. However, it does not provide a good design for the construction of a practical heat engine, and in practice, other cycles are used. These include the **Brayton Cycle** for gas turbines, the **Diesel** and **Otto cycles** used in engines for land-based transportation, as well as the **Rankine cycle** for steam turbines. Figure 5.6 shows the ideal power cycles for the Diesel and Rankine cycles. Both the *P–V* and the *T–S* plots for the Diesel and Rankine cycles are shown.

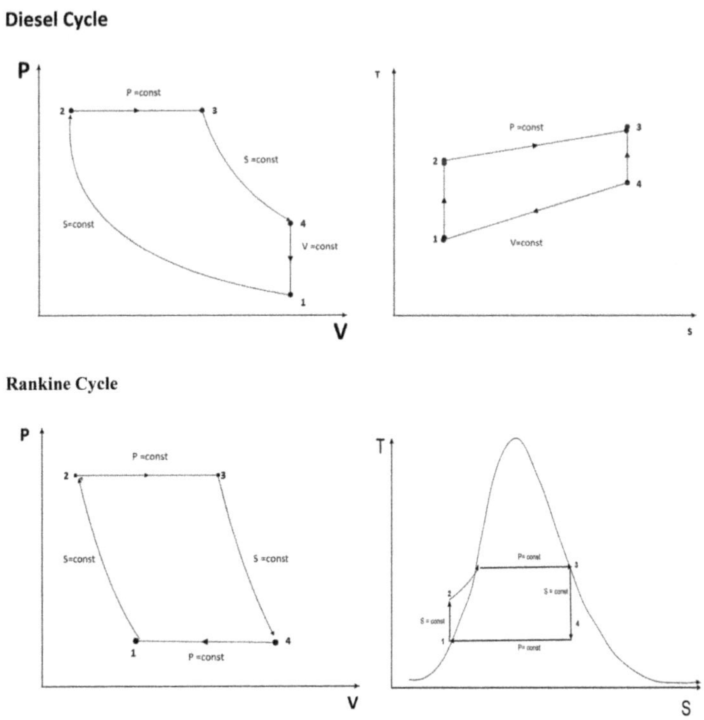

Figure 5.6. Ideal power cycles related to the diesel engine and steam turbine (both *P–V* and *T–S* state diagrams are included for each cycle).

5.5.1 Closed and open cycles

The discussion so far has assumed that the working fluid is taken through a cyclic process in all heat engines. This is not quite true, as some heat engines allow (at least part of) the working fluid to exit the engine at some point. (The theory of heat engines developed above can still be applied to such engines.) The power cycles of heat engines can thus be divided into **closed cycles** and **open cycles**.

A closed cycle is one where the working fluid is taken through the full cyclic process and returned (for recycling) to its original position in the engine. The working fluid ordinarily remains inside a sealed section inside the engine. An open cycle is one where the working fluid is taken through the full cyclic process but exhausted from the engine at the end of the cycle. All engines with an exhaust (tailpipe) are of this type.

5.6 Types of heat engines

There are several types of heat engines possible, depending on their design and the manner in which they convert heat energy into mechanical energy (i.e. do work on the external environment). Heat engines can be either **external combustion engines** or **internal combustion engines**. They can use either turbines or **piston-cylinder systems** to convert heat into work.

5.6.1 External and internal combustion engines

An **external combustion engine** is one where the fuel is **burnt outside the engine**. The burning usually takes place inside a chamber called variously a **furnace, firebox, boiler or combustor**. Examples of such engines are the **steam engine** and the **Stirling engine**. In such engines, the working fluid remains sealed inside a system of tubes/cylinders connected to other components, and is heated by conduction/convection as it passes through the furnace (see figure 5.2).

An **internal combustion engine** (abbreviated to **ICE** or **IC engine**) is an engine where the fuel is **burnt inside the engine** itself. The most common examples are **petrol or gasoline engines** (also called **spark ignition (SI) engines**), and **diesel engines** (known also as **compression ignition (CI) engines**) used in land-based transportation.

5.6.2 Turbines and piston engines

The heat engine usually converts heat into mechanical energy through two mechanisms: **turbines** and **pistons**. In a turbine engine, the working fluid is expanded in an axle-mounted device, called a **turbine**, that is caused to turn (i.e. rotate) when the working fluid is expanded through it (as in figure 5.2).

In internal combustion engines (ICE), it is most common to use a **piston-cylinder arrangement** (as depicted schematically in figure 5.3). The fuel is burnt inside the cylinder and causes the piston to go through oscillatory motion that is translated into the turning action of an axle. Engines based on this principle are called **reciprocation engines**. Most common diesel and gasoline (petrol) fuelled road transportation vehicles use internal combustion engines that are reciprocating engines. One must note that it is also possible to design external combustion engines that use the piston principle to convert heat into work (see e.g. the Stirling engine.)

(See quiz 5.1—Types of heat engines.)

5.7 The Rankine cycle and the steam turbine

The sections above have elaborated on how heat engines work, the basic physics of how heat energy is converted into work, and the various types of heat engines that are commonly used.

It is now time to consider the practical aspects of a real heat engine in more detail. One can begin by choosing a specific power cycle. An appropriate choice is the Rankine cycle, which provides the basis of operation of the **steam engine** (also called the steam turbine). This is one of the most common heat engines used for power generation, and provides the central power generating component (sometimes called the ***prime mover***) for thermal power plants.

Before proceeding with the analysis of heat engines, it is instructive to consider the nature of working fluids of heat engines in more detail. The main purpose of the working fluid is to convert heat energy that enters the engine into useful work. This process involves heat changes in the working fluid, which may in turn lead to a change of phase of the medium. For instance, heating a liquid may change its phase to the gaseous phase during the full cycle of this working fluid. This will introduce considerations of the latent heat of vaporisation in the analysis.

So far, it was assumed that the phase of the working fluid does not change during a power cycle, making it unnecessary to take the phase of the working fluid into account when describing the cyclic process. This, however, is not always true. In the case of the Rankine cycle for instance, the working fluid (water) changes from a liquid to a liquid–vapour mixture, and then to pure vapour (steam) before it is returned to the pure liquid phase during the course of the cycle.

These changes cannot be fully described in terms of state diagrams alone. To include information relating to the phase of the working fluid, one needs to make use of **property diagrams**. These provide information on the **phase** as well as the working cycle of the fluid. Such diagrams divide the state diagram into regions in which the phases of the fluid are different, ranging from liquid to liquid–vapour mixture to pure vapour. These regions are demarcated by a dome-shaped curve called the **saturation dome**. Such **property diagrams** provide the minimum information required for an adequate analysis of the behaviour of the working fluid during the working cycle.

The property diagram for the ideal Rankine cycle of a steam engine is shown schematically in figure 5.7.

In keeping with the state diagrams discussed so far, this figure displays the four processes of the ideal Rankine cycle (as presented in the P–V plot of figure 5.6). But it goes further to provide information on the phase of the working fluid by dividing the area of the P–V plot into three distinct regions by the dome-shaped curve called the **saturation dome**.

To the left of the dome is the **Sub-Cooled Liquid** zone, where the working fluid (water in this case) is in the liquid state. This zone is separated on the right from the

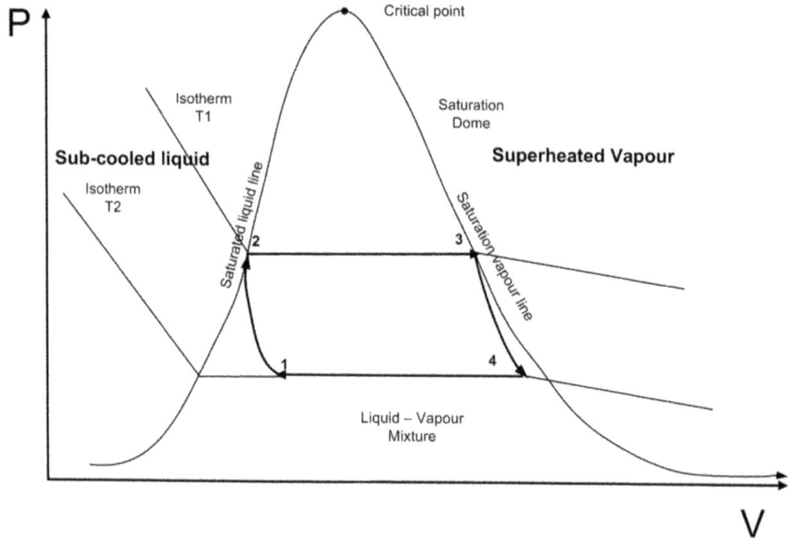

Figure 5.7. A property diagram (shown as the *P–V* plot) for the ideal Rankine cycle for steam engines. Such diagrams include information on the phase of the working fluid as well as the working cycle.

Liquid–Vapour Mixture zone, where the fluid exists as a mixture of liquid and vapour (steam) phases, separated by the **Saturated Liquid Line**. This line demarcates the physical conditions (as described by the state variables) necessary for the fluid to start entering the vapour phase (i.e. evaporating) as one moves from left to right in the diagram.

To the right of the dome is the **Super-Heated Vapour** zone. Here the conditions (described by the values of the state variables) enable the fluid to remain totally in the vapour phase. This zone is separated from the mixed zone on the left by the **Saturated Vapour** line. The phase of the fluid begins to turn from vapour to liquid as one moves towards the left from this line. The two saturated lines meet at the **Critical Point** at the top of the zone.

An additional feature of the property diagram is the presence of lines along which a particular state variable has a constant value. Note that whereas the complete description of the state of a system ideally requires the use of the four state variables, only two of these are explicitly referred to in state diagrams. Property diagrams include reference to a third variable through the use of lines of constant value. In figure 5.7, the temperature variable is included by showing lines of equal temperature, called **isotherms**. In the property diagram of the *T–S* plot, the pressure is included in a similar way by the use of **isobars** (which are lines of equal pressure).

5.7.1 The realistic steam turbine with super-heating

Examination of the property diagram shown in figure 5.7 reveals practical issues that will arise if the ideal Rankine cycle is used as the power cycle for the operation of a steam turbine. The diagram shows that both states 1 and 4 of the system will lie in the liquid–vapour mixture region. This presents two issues.

Firstly, as the pumps used to compress fluids work better on liquids, it is more productive to use a cycle where state 1 of the system is moved to the saturated liquid line or beyond. Secondly, as water droplets cause damage to turbine surfaces, state 3 of the system is best moved up and out of the dome altogether into the super-heated vapour region. Implementing such changes also increases the thermal efficiency of the heat engine. Figure 5.8 shows the resulting, practical cycle on a *T–S* plot.

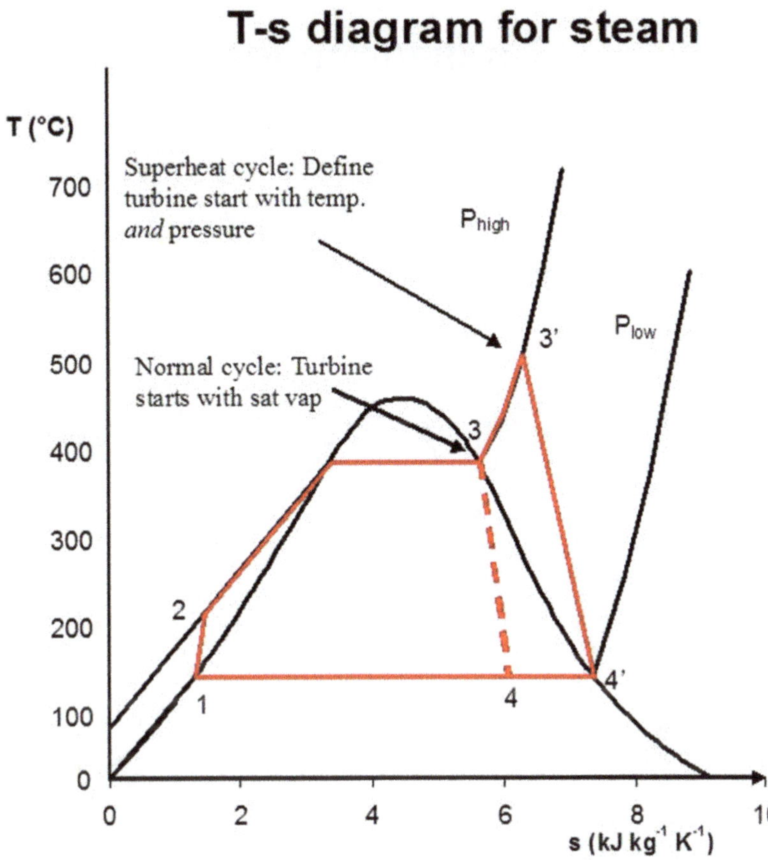

Figure 5.8. Property diagram (*T–S* plot) of a realistic steam turbine with super-heating. This Rankine cycle with superheating has been obtained by the author from the Wikimedia website where it was made available under a CC BY-SA 3.0 licence. It is included within this article on that basis. It is attributed to Donebythesecondlaw.

(See exercise 5.5—Super-heating and thermal efficiency.)

5.7.2 Improving design with reheating

The overall performance of the heat engine can also be improved by adopting a new design where two turbines are used. In this design, the exhaust working fluid from the first (high pressure) turbine is passed back into the boiler for reheating, and emerges to enter the second turbine at a lower pressure. The net result is that while the reheated temperature remains close to the temperature at the inlet of the first turbine, the pressure at the inlet of the second turbine reduces to a much lower value.

The overall effect is the avoidance of droplet formation and the need to go to very high temperatures (e.g. close to or exceeding ~600 °C, the creep temperature of steel). The property diagram for a steam engine with reheating is shown in figure 5.9.

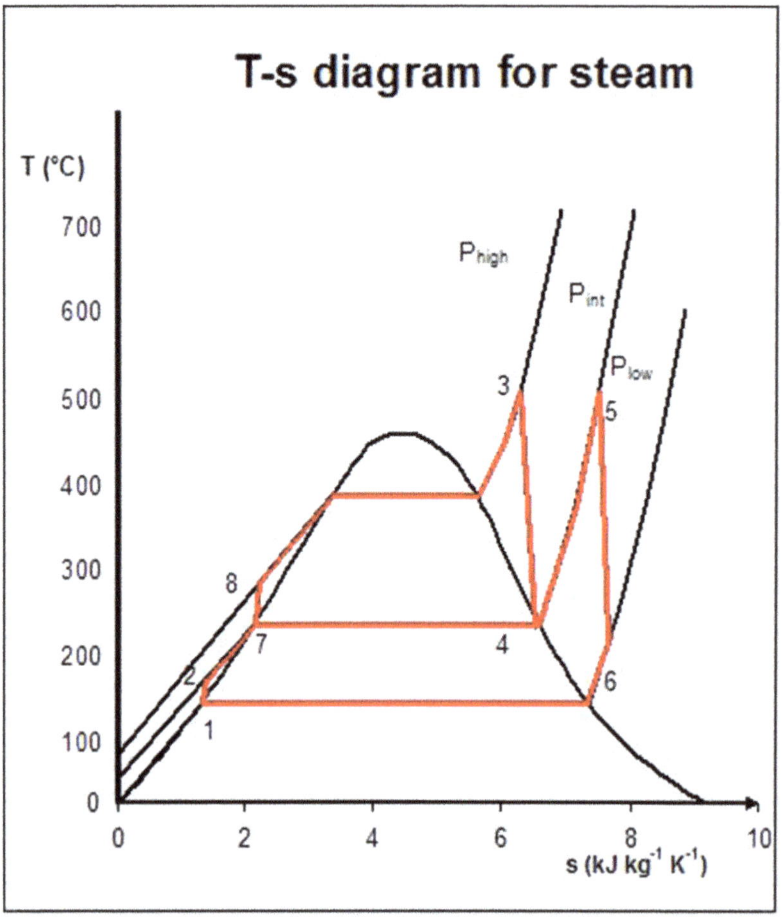

Figure 5.9. The property diagram of a steam turbine with reheating [6]. This regenerative Rankine cycle has been obtained by the author from the Wikimedia website where it was made available under a CC BY-SA 3.0 licence. It is included within this article on that basis. It is attributed to Donebythesecondlaw.

5.8 The Brayton cycle and the gas turbine

Another heat engine that is of considerable relevance to transportation and power generation is the **gas turbine**. Gas turbines provide the power source for a large portion of both the transportation and stationary energy use sectors. They are used to power aircrafts, ships, trains, military tanks as well generating electricity in power plants.

Gas turbines operate on the principle of the **Brayton cycle**, with air as the working fluid. Two important features to note about gas turbines are firstly that they are open cycle devices, and secondly, they do not involve a phase change in the power cycle.

In the Rankine cycle-operated steam turbine, the working fluid changes from a liquid to a vapour (steam) via a mixed liquid–vapour phase in the working cycle of the engine. In the case of gas turbines however, the fluid remains gaseous throughout the cycle (thus avoiding the need to address design issues). In addition, as the working fluid is exhausted to the atmosphere at the end of the working cycle, the gas turbine does not need to employ the final process of the ideal Brayton cycle.

The essential components of a gas turbine consist of a **compressor**, a **combustion chamber** and a **turbine**, with the compressor and the turbine mounted on the same axle. During operation, the compressor compresses the incoming air from the atmosphere to high pressures. This compressed air is then led to the combustion chamber, where it is mixed with injected fuel and the mixture burnt.

The hot gas at high pressure that results is expanded in the turbine, which turns and provides the thrust for either a moving vehicle such as an aircraft, or the shaft power for turning an electrical generator in a power station, depending on the application. Because of the axial coupling of the turbine with the compressor, the latter is automatically provided the torque required for it to produce the gas compression at the initial stage of the cycle.

The ideal Brayton cycle consists of four processes, comprising an isentropic compression, an isobaric heat input, an isentropic expansion of the working fluid and an isobaric heat rejection.

The operation of the gas turbine may be associated with the ideal Brayton cycle as follows:

- Process 1—isentropic (i.e. adiabatic) compression of air in the compressor.
- Process 2—isobaric addition of heat via the combustion of the air–fuel mixture in the combustion chamber.
- Process 3—isentropic expansion of high temperature gas through a turbine.

The fourth process expected from the Brayton cycle, namely the isobaric rejection of heat, is not used in gas turbines as they are open cycle devices, and reject the working fluid to the atmosphere rather than recycling it.

An issue with this simple mode of use of the gas turbine is that the gas is ejected from the system at a high temperature, thus taking away a large fraction of the energy input from the fuel. This naturally leads to low turbine efficiency. Some of this lost energy can be recovered if the hot gas is used to pre-heat the compressed gas before it enters the combustion chamber (thereby requiring less fuel to heat the compressed air entering the turbine). This heat exchange between the hot exhaust gas and the incoming compressed gas can take place in a **regenerator**, which is inserted into the operational cycle between the compressor and the combustion chamber. Figure 5.10 shows a schematic depiction of a gas turbine engine that includes a regenerator.

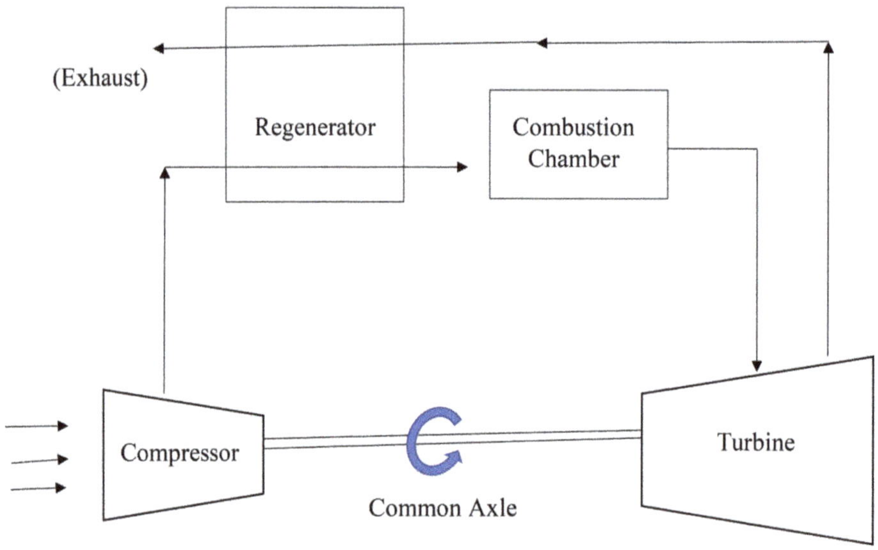

Figure 5.10. Basic design of a regenerative gas turbine used in gas turbine power plants (source: after [7]).

(See research question 5.1.)

Exercises, questions and quizzes

Exercises

Exercise 5.1. States and paths.

On a P–V diagram, draw the path that takes the system from an initial state to a final state by:

(i) an isobaric process;
(ii) an isochoric process;
(iii) an isentropic process;

starting from the same initial state. (*Hint:* Assume the system is an ideal gas.)

Exercise 5.2. Work done on a system.

a) Show that work is done on a system when its volume is decreased by the application of pressure.
b) Show that the work done by a system when it is taken from an initial state to a final state is equal to the area under its P vs V curve that describes the path for the process.

Exercise 5.3. Thermal efficiency of a heat engine.

a) Estimate the heat rejected to the cold reservoir per cycle of a heat engine that is running at 1000 rpm, if 0.001 L of fuel (which has a calorific value of 30 MJ L^{-1}) is injected into it per cycle. Assume that the thermal efficiency of the engine is 40%.

b) State an assumption you needed to make in arriving at your answer.
c) What are two possible ways in which you could improve the efficiency of this engine?

Exercise 5.4. Deducing paths of processes.

If the working fluid is an ideal gas, draw the shapes of the paths that the system will take on a *P*–*V* diagram for an isothermal and an isochoric process as the volume increases. Start the two paths from a common initial state i.

Exercise 5.5. Super-heating and thermal efficiency.

Compare the *T*–*S* plot of the ideal Rankine cycle shown in figure 5.6 to the realistic cycle shown in figure 5.8.

a) What does this indicate about the relative amounts of work output delivered in the super-heated and the ideal cases?
b) What does this imply about the relative efficiencies of the two arrangements?

Questions

Research question 5.1. Aircraft jet engines and gas turbine power plants.

As noted in section 5.8, gas turbines are used for both moving (transportation vehicles) and stationary (power plant) devices. In transportation, they are particularly important in aircraft, where five variations (turboprop, turbojet, turboshaft, turbofan and ramjet) of the basic gas turbine are used as engines. The gas turbine also provides the prime mover for the gas turbine (GT) power plant.

In applications such as transportation which involves the translational motion of vehicles, the gas turbine must produce a thrust (i.e. a force on the vehicle that drives it in the forward direction). According to Newton's third law, the thrust must be a reaction force. This can be a force that is produced either by ejecting mass (in the form of a hot gas) or the action of a rotating propeller against atmospheric air.

In power plants, translational motion is not required, and the only requirement is for the gas turbine to convert heat energy into electricity. This is achieved by producing shaft power that drives an electrical generator.

a) Describe the principle of operation of an aircraft jet engine, and show how it is able to produce and maximise the motion of an aircraft.
b) Briefly outline the differences between the five variations of the basic turbine as jet engines for aircrafts.
c) Compare the operation of an aircraft jet engine to that of the gas turbine in a GT power plant. Describe in detail the principle of operation of this latter engine, clearly showing how it differs from that of a jet engine.
d) Explain why the use of a regenerator improves the efficiency of GT power plants.

Quiz

Quiz 5.1. Types of heat engines.

1. What is the difference between an external combustion engine and an internal combustion engine?

2. Differentiate the following heat engines into external and internal combustion engines:
    a) A steam engine used in electrical power generation
    b) The engine of a gasoline-fuelled vehicle
    c) A diesel engine for an electrical power plant
    d) The steam-driven engine of an early locomotive train.
3. In a heat engine, the device that turns as the working fluid (usually steam) passes through it is called:
    a piston, (b) a cylinder, (c) a turbine, (d) none of these.
4. A diesel engine is an example of a
    reciprocating engine, (b) steam engine, (c) steam turbine.
5. A Stirling engine is:
    an IC engine, (b) an external combustion engine, (c) a rotary engine, (d) none of these.

References

[1] Thermal Engineering. *What is Fourier's Law of Thermal Conduction* [accessed 8 November 2022]. Available from https://thermal-engineering.org/what-is-fouriers-law-of-thermal-conduction-definition/

[2] Rankine cycle. Wikipedia. 2007 [accessed 11 November 2022]. Available from https://en.wikipedia.org/wiki/Rankine_cycle

[3] *What Are PV Diagrams?* (Mountain View, CA: Khan Academy) [accessed 11 November 2022]. Available from https://khanacademy.org/science/physics/thermodynamics/laws-of-thermodynamics/a/what-are-pv-diagrams

[4] See, e.g. ed Serway R A and Jewett J W 2018 *Physics for Scientists and Engineers* 10th edn (Boston, MA: Cengage)

[5] See e.g. ed Cengal Y and Boles M 2014 *Thermodynamics: An Engineering Approach* 8th edn (New York: McGraw-Hill Education)

[6] Rankine cycle. Wikipedia. 2007 [accessed 11 November 2022]. Available from https://en.wikipedia.org/wiki/Rankine_cycle. Attribution: Donebythesecondlaw.

[7] Regeneration is gas turbine. Mech Content [accessed 11 November 2022]. Available from https://mechcontent.com/regeneration-gas-turbine/

IOP Publishing

Bioenergy for Power Generation, Transportation and Climate Change Mitigation

Anirudh Singh

Chapter 6

Bioenergy-based power generation technology

Thermal power plants use heat engines as their main component (called the **prime mover**), and differ essentially in the fuel they use and their fuel-handling facilities. They may be divided into condensing or combined heat and power (CHP) plants. To improve their efficiencies, the heat engines may be combined to produce combined cycle power plants.

This chapter examines the principle of operation of the typical thermal power plant, and elaborates on the various types of such plants that are in use today. It ends by considering the range of bioenergy-fired thermal power plants in existence today, most of which have been in industrial application for decades, and providing a case study of a (waste-to-energy) biogas CHP in operation in Hamburg, Germany.

6.1 Introduction

The last chapter introduced the physics and technology of bioenergy power by providing an introduction to the basic thermodynamics of the heat engine and its application to the operation of steam engines and gas turbines. This chapter goes on to consider the implementation of these principles in actual thermal power plants.

The chapter begins by examining the generic design of all thermal power plants, which must include a heat engine as the **prime mover** and an energy source as well as an electrical distribution system. The types of power plants, including their division into the two major categories of **condensing** and **combined heat and power (CHP)** plants are discussed in section 6.3.

Power plants may use heat engines which are combined in tandem so that the heat lost from one stage provides the input energy to the next stage. Additionally, energy sources such as gasifiers may be linked directly to heat engines. Such combinations are discussed in section 6.4.

doi:10.1088/978-0-7503-3555-3ch6

Finally, section 6.5 provides an overview of the range of bioenergy-fired thermal power plants that are in operation today. They include the **bagasse-fired CHPs** that have been used by the sugar industry for many decades, **solid biomass-fired power plants**, **pyrolysis oil powered plants** and **biogas power plants**. A case study of a biogas CHP in Germany serves as an example of the capability of biogas plants that generate their fuel from commercial and domestic waste, and provide district heating in addition to their main function of electricity generation.

6.2 Principles of operation of thermal power plants

Thermal power plants are installations that use heat to produce electrical energy. A mandatory requirement is therefore the employment of heat engines as their main functional unit (often called the **prime mover**). As a consequence, their outputs and maximum performances are subject to the first and second laws of thermodynamics and limited by Carnot efficiency.

6.2.1 Generic design of thermal power plants

There are a wide variety of thermal power plants, varying in form and physical design, that meet these requirements. However, they all share the same basic functional features (i.e. have the same generic design). All thermal power plants must have:

(i) an energy source;
(ii) a heat engine; and
(iii) a balance of plant that includes some sort of fuel-handling and power distribution capability.

Indeed, once the energy source has been determined, the rest of the thermal power plant is essentially the same for most types of installations. Figure 6.1 is a diagrammatical illustration of the generic features shared by all thermal power plants.

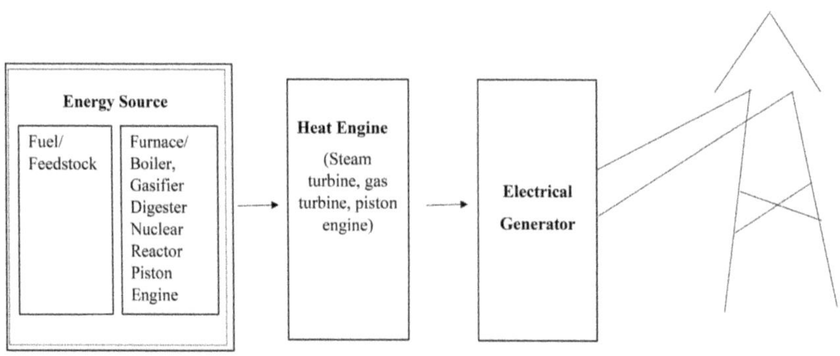

Figure 6.1. The essential features of a thermal power plant. All thermal power plants share these common features.

6.3 Types of thermal power plants

Thermal power plants may be either **condensing power plants** that produce only power, or **combined heat and power (CHP) plants** that produce both heat and power. The latter may consist of back-pressure power plants, or (the more efficient) extraction-condensing plants.

Most power plants today employ heat engines that use the Rankine cycle for their operation. But others may use gas turbines as their heat engines. These latter engines are also used to provide propulsion for jet aircraft in the aviation industry. This section provides an introduction to all such uses of the heat engine to produce power and propulsion.

6.3.1 Characteristics of thermal power plants

The differences that characterise individual types of power plants arise mainly from the **nature of the energy source**. Another important difference that must be noted however, is related to the **intended use of the power plant**. That is, whether it is to be used for generating electricity alone or for the purpose of both electricity generation and provision of process heat.

The energy source can be wide ranging, including **nuclear**, **geothermal**, **fossil fuel** (including **coal**, **oil** or **gas**), or **bioenergy sources** such as woody biomass, biomass residues, sewage waste and MSW (see table 2.2). The specific features of the power plant will depend on the nature and physical phase of the energy source.

In spite of the diversity in their energy source, thermal power plants are strikingly similar in other features. For instance, a power plant using atomic (fission) energy as the energy source will require a nuclear reactor, a heat engine, an electrical generator and balance of plant to produce electricity. A coal-fired power plant will need coal-handling facilities, and a furnace to burn the fuel, but will otherwise have a similar complement of remaining components as the nuclear power plant. If a biogas-fired power plant is considered, then the steam engine will need to be replaced either with a gas turbine or a reciprocating engine capable of running on gas.

(See exercise 6.1—Types of thermal power plants.)

6.3.2 Condensing and CHP power plants

As noted above, apart from differences in design due to the fuel used, the physical characteristics of a thermal power plant will also depend on whether it is to be used as a dedicated electrical power generation facility, or whether it is to have the dual purpose of generating electricity as well as supplying process heat (for industrial and other needs).

A **condensing power plant** is one that is dedicated solely to electrical power generation. To maximise their efficiency, such power plants need to reduce the temperature of the working fluid after it passes through the turbine. This can be best seen in terms of the efficiency of the ideal heat engine (see equation (5.9)).

The condensing power plant may use **cooling towers** or water from a large water source such as a river or lake nearby to provide the cooling. For this reason, power plants are frequently supplied with cooling tower constructions, or are situated beside water sources such as running rivers or large lakes.

A co-generation, or **combined heat and power plant (CHP)** is one that produces process heat for industries as well as electricity. The efficiency of such machines is determined by the total amount of incoming heat that is utilised for useful purposes. This includes heat used for industrial processes as well as electrical power generation.

The process heat is usually derived from the spent steam exhausted by the power generation stage. As a result, the efficiency of a CHP plant is much higher than that of a condensing power plant. CHPs are quite popular in industrial applications, and examples of their use will be discussed in section 6.5. The possible designs of CHP power plants are elaborated on further in box 6.1.

Box 6.1. The difference between condensing and CHP power plants.

Because the sole output of the **condensing power plant** is electricity, special effort must be made to increase its efficiency by maximising the furnace temperature and minimising the temperature of the spent steam before it is recycled. In contrast, the temperatures in the **combined heat and power plant (CHP)** will be determined to a large extent by the ratio between the electrical and thermal energy it is required to generate for its needs.

Condensing power plant

The essential features that a **condensing power plant** must possess to achieve its aim of power production are a **furnace/boiler**, where the fuel is burnt to produce the heat as input into the system, a **turbine** to convert heat into mechanical energy, a **condenser** to condense the spent steam back into liquid water, and a **feed pump** to return the condensed working fluid back to the boiler.

Figure 5.2 in chapter 5 provides a simple schematic diagram that depicts most of these processes. The condenser at the right of the diagram is usually connected to a cooling system that uses either a river or cooling towers for the cooling. The actual plant is characterised by the conspicuous presence of cooling towers (large cylindrical structures) and/or a river or large lake nearby to supply water for cooling the steam exhausted from the turbine. The coal-powered power plants to be discussed below are usually condensing power plants.

Combined heat and power plant (CHP)

The CHP produces both heat and electrical power as the desired outputs. These have been used in the sugar and timber industries for nearly a century [1]. The two variations of the CHP steam engine used in the sugar industry are:

- the back-pressure (or non-condensing) steam turbine system, and
- the condensing-extraction steam turbine.

Back-pressure turbine systems

Back-pressure turbine systems were among the first CHP systems used in the sugar industry since the earliest days to provide the power and process heat requirements of sugar mills [2, 3].

A schematic diagram of such a system is shown in figure 6.2.

Note that this turbine system:

- is directly connected to an electrical generator that produces electricity, and
- does not have a condenser that condenses the spent steam from the turbine.

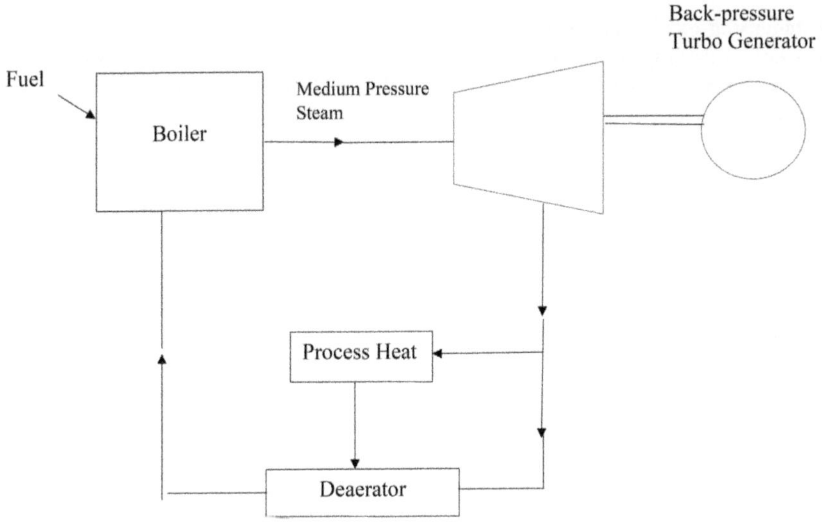

Figure 6.2. A back-pressure CHP steam turbine used in traditional sugar mills (after [4]).

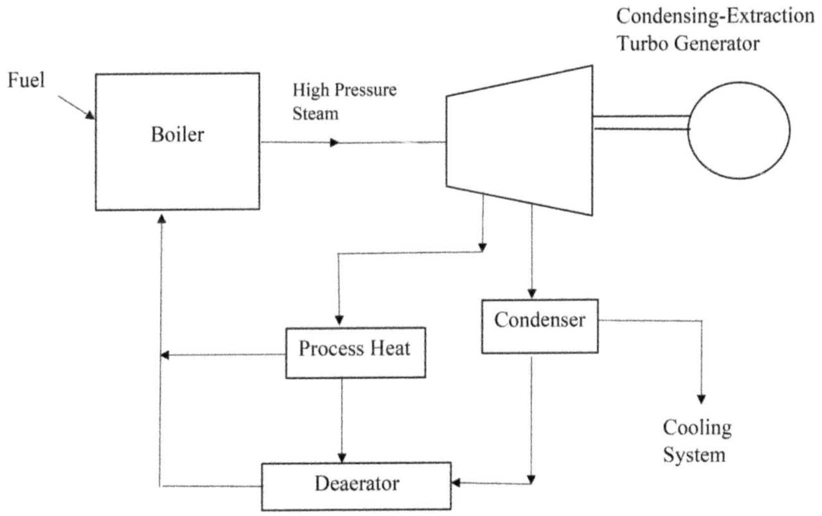

Figure 6.3. A schematic diagram of a (high efficiency) extraction-condensing CHP steam turbine, which is capable of exporting power to the grid (after [4]).

These turbine systems were used to provide the basic process heat and in-house electricity requirements for sugar mills from the earliest days of the industry [2, 4]. They employed low pressure and temperature boilers, (typically 24 bar and 300 °C [4]), which had low efficiencies. These back-pressure CHPs were operated during sugarcane crushing seasons only. The low efficiencies sufficed for the needs of the sugar mill as there was no requirement for exporting electricity to the grid.

Condensing-extraction turbines

Recent mills employ more efficient steam generation systems that are capable of producing excess power for export to the grid [4]. These **condensing-extraction** turbine systems are efficient enough to be able to produce and export electricity that is surplus to the needs of the sugar mill. Figure 6.3 shows a schematic diagram of such a turbine system that provides the prime mover for these modern CHP plants.

In this turbine system, a fraction of the high-pressure steam is extracted from the turbine for the process heat requirements of the mill, while the remainder drives the turbine and is condensed after exiting from it as in the case of condensing turbines.

Such CHP plants can operate 365 days/year using stored bagasse and coal/wood chips as supplementary fuel. The pressures/temperatures utilised are typically in excess of 60 bars and 500 °C, allowing for greater plant efficiencies and better utilisation of bagasse.

6.3.3 A typical condensing coal-fired power plant

Because of their similarities, it is possible to obtain a good insight into the way all thermal power plants operate by considering a typical example. Such an example is provided by a coal-fired power plant. The basic features of such a system is displayed in figure 6.4.

This typically consists of a **coal-handling/processing facility**, an **air-handling facility**, a **boiler (also called a furnace)** section, a **multi-stage turbine**, a **condenser system**, as well as the **electrical generator** system for power generation and distribution (see figure 6.4). The electrical power production path starts from the coal-yard (bunker) where the coal is crushed and pulverised in the mill and fed into the furnace/boiler.

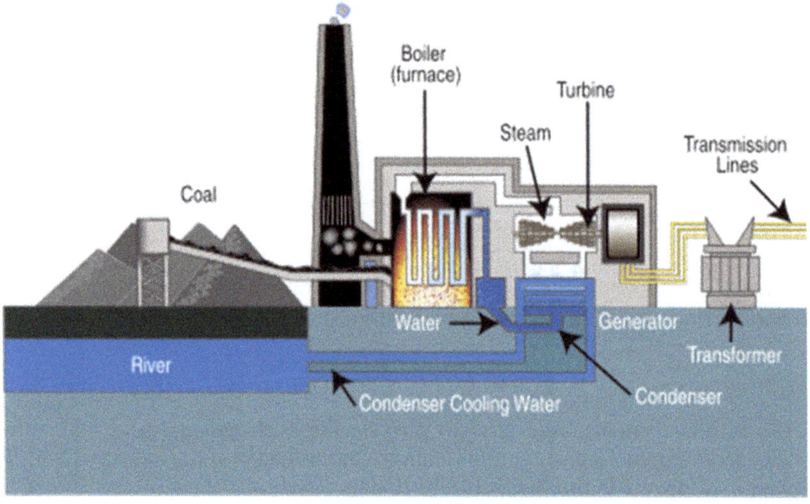

Figure 6.4. A schematic diagram of a typical coal-fired thermal power plant indicating the fuel handling, heat input, electrical power generation and the cooling mechanism for the working fluid. Reproduced from [5]. Image stated to be in public domain. https://www.tva.com/.

A typical power plant may contain as many as three turbines (accepting steam at high, intermediate and lower pressures), and the power production cycle may be envisaged as follows.

In the boiler, the (de-mineralised) water is boiled to produce super-heated steam which is then fed to the high-pressure segment of the turbine (HPT). A portion of the spent steam from the HPT is sent to the boiler for reheating, and delivery to the intermediate pressure turbine (IPT). The lower-pressure steam produced by the IPT is sent to the low pressure turbine (LPT) that provides the main generation capacity of the power plant. The three turbines are mounted on a common axle that they share with the electrical generator.

The spent steam from the LPT is condensed in the condenser which is coupled to a cooling tower and/or a large lake/river system. These provide alternative means for reducing the temperature of the working fluid (water) as far as possible. The solid effluent from the boiler, consisting of the fly ash component of the combustion products, is first passed through an electrostatic precipitator (ESP) which removes the dust particles, allowing the hot gas to be released to the atmosphere via the stack (also called a 'chimney').

(See exercise 6.2—The thermodynamics of a thermal power plant.)

6.4 Combined cycle power plants

6.4.1 Multi-stage power plants

The ability of a thermal power plant to convert the chemical energy of the fuel into useful work depends on the efficiency of the heat engine it uses as its prime mover. The expression for the thermal efficiency of such an engine is given by expression (5.8) and that for an ideal (Carnot) engine by expression (5.9).

It is clear from the second law of thermodynamics that some heat (Q_c) has to be rejected for the engine to work, as no heat engine can be 100% efficient. But is it possible to re-use this rejected heat from the engine somehow?

This is indeed possible, as the second law does not prevent us from re-using the heat rejected from one heat engine as the input heat for another steam engine. One can increase the total efficiency of a thermal power plant by combining two heat engines in tandem, so that exhaust steam of one provides the input energy for another. A power plant that does this is called a **combined cycle power plant (CCPP)**. The total efficiency of such combinations of heat engines is given simply by the product of the individual efficiencies.

(See exercise 6.3—Efficiency of combined cycle power plants.)

As noted in the last chapter, the temperatures at the exhaust of gas turbines are quite high, leading to low turbine efficiency. The efficiency can be improved by combining the gas turbine with a steam turbine that uses the exhaust of the gas turbine as input.

6.4.2 Integrated gasification combined cycle power plant (IGCCPP)

It was noted in chapter 3 that biomass could be turned into syngas using a gasifier. While this fuel may be used to power thermal power plants, its manufacture and transportation to the plant is inconvenient and cost-incurring. Nonetheless, syngas can still be used to produce electrical power using a gas turbine in a specially-designed power plant. This can be achieved by combining the gasifier directly to a gas turbine to produce an **integrated gasification combined cycle power plant (IGCCPP)**.

(See question 6.1—Power output of an IGGPP.)

6.5 Bioenergy-fired thermal power plants in industrial application

The previous sections have considered the basic technology of thermal power generation. We now consider the implementation of such technologies to the specific case where the energy source is bioenergy. This form of energy can occur as solid, liquid or gaseous fuel. The associated technologies available today are either mature technologies or at the developmental stage.

Two cases of power generation involving solid biomass are bagasse-fired power generation by the sugar industry, and the use of woody biomass obtained either from forestry residues or dedicated energy farms. A liquid form of bioenergy that has been undergoing progressive development over the last two decades is pyrolysis oil (also called bio-oil), which is emerging as a promising new fuel for the power industry. In addition to these solid and liquid fuels, biogas, derived from a range of industrial and domestic wastes, is also proving to be a viable feedstock for a significant fraction of the power generation industry in the northern hemisphere.

We begin by considering how the sugarcane industry has been using one of its main by-products for power generation and process heat requirements throughout its recorded history.

6.5.1 Bagasse-fired power plants in the sugar industry

Bagasse, the ligno-cellulosic by-product produced by the sugar industry during the crushing of the sugarcane stalk, provides a readily-available bioenergy source to this industry sector for power generation and process heat. From its earliest days, the industry has used this fuel source for its co-generation (i.e. CHP) power plants. Mauritius and Hawaii were the first to use co-generation in the sugar mills. According to one report, in 1926/1927, significant proportions of the electricity needs of these countries (26% in the case of Mauritius and 10% for Hawaii) were provided by sugar mills.

An overview of how co-generation has been utilised in leading sugarcane producing countries is presented by Birru and others [6]. Their work may be summarised briefly as follows.

6.5.1.1 Brazil

Brazil is the largest sugar producing country in the world. In 2012, its 400 sugar mills produced 590 million tons of sugarcane, yielding 40.2 million tons of raw sugar and 23 million litres of ethanol. In 2009, the industry generated 4.6 GW power, of which 25% was sold to the national grid.

6.5.1.2 India

India is the second largest sugarcane producer in the world. In 2013/2014, its 509 mills produced 2300 MW of power, of which 1300 MW was exported to the grid.

6.5.1.3 Australia

Australia produces most of its sugarcane in Queensland. In 2012, the country's 24 sugar mills exported more than half of their total electricity generation (of 400 GWh) to the grid.

6.5.1.4 Mauritius

Mauritius was the first country (in 1957) to commence producing electricity for grid supply from its sugar mills. In 2012, Mauritius produced 3.9 million tons of cane. All sugar mills in the country have now been upgraded to produce surplus power. The co-generation systems operate at steam cycles of 82 bars and 525 °C, producing 75–140 kWh per tonne of cane of surplus electricity. The mills burn coal during off-seasons to continue supplying power to the grid.

6.5.2 Solid biomass-fired power plants

Solid biomass feedstock for power generation can be either derived directly from plant sources, or obtained as solid material derived from mixed sources such as food industry residues, animal wastes, urban wastes, municipal solid wastes and sewage wastes (see table 2.2 for a comprehensive list of such energy resources).

Plants that use such feedstock are described as either **thermal power plants** or **incineration power plants**, with the latter designation applied to the waste products (and often described as waste-to-energy (WTE) power generation).

The power plant architecture of solid biomass-fired power plants is generally very similar to that of a coal-fired thermal power plant described schematically in figure 6.4, with the main difference lying in the feedstock handling section. This differs slightly between plants using feedstock derived directly from plant material, and those using the organic fractions of municipal solid waste (MSW). Both require some degree of chop-length processing and/or densification. The latter will almost certainly require some pre-treatment and emissions reductions/monitoring to satisfy the health regulations of the country.

(See research question 6.1—Plant-derived biomass and MSW as fuels for power plants.)

6.5.3 Pyrolysis power plants

As seen in chapter 3, the main products of pyrolysis are pyrolysis oil (also called bio-oil), (solid) bio-char and non-condensable gases. Pyrolysis oil forms the major component of products (60–70%) in the temperature range of 450–500 °C.

While research in the pyrolysis (and notably fast pyrolysis) of biomass has progressed steadily over the last several decades, it was not until early last decade that serious efforts were made in the commercialisation of pyrolysis at the industrial

Table 6. 1. Commercial fast pyrolysis plants and their status in 2020 (data source: [10]).

Company	Country	Reactor technology	Bio-oil production capacity (kg bio-oil/h)	Status/year
Pyrocell (BTG Bioliquids)	Sweden	BTG-BTL rotating cone	3250	Operational/ 2021
Green Fuel Nordic (BTG Bioliquids)	Finland	BTG-BTL rotating cone	3250	Operational/ 2022
Twence/Empyro (BTG Bioliquids)	Netherlands	BTG-BTL rotating cone	3250	Operational/ 2015
Savon Voima Joensuu	Finland	VTT Fluid bed/ riser	6500	Operational/ 2013
AE Cote-Nord Bioenergy	Canada	Ensyn Fluid bed/riser	6400	Commissioning/ 2017
Red Arrows	Canada	Ensyn Fluid bed/riser	—	Operational/ 1996
Kerry Group plc	Canada	Ensyn Fluid bed/riser	—	Operational
Genting	Malaysia	BTG-Rotating cone	—	Dormant/2014
KiOR	USA	Catalytic fast pyrolysis	—	Dormant/2014

scale for various applications. The product of main interest was pyrolysis oil, which is usually obtained through a fast pyrolysis technique.

This product has several potential uses. According to BTG Bioliquids, bio-oil can be used to produce heat, sustainable transport fuels and bio-based chemicals [7]. Indeed, according to this source, this pyrolysis product can replace crude oil in all the mentioned sectors.

The centre of interest in the industrial application of bio-oils appears to be in the EU countries, (in particular Finland), who are currently the main players in the technological application of pyrolysis products. Fortum/Valmet, a Finnish company, was the first to commercialise bio-oil production when it commissioned an industrial-scale fast pyrolysis bio-oil plant at Joensuu in East Finland on 29 November 2013 [8]. This plant was integrated with a CHP plant already in existence at the Joensuu site.

The commercialisation of bio-oil received a major boost in 2014 when Fortum, UPM and Valmet formed a consortium on a five-year project called ***LignoCat*** to develop and commercialise integrated catalytic technology to produce biofuels from ligno-cellulosic feedstock such as forestry residues [9].

There has been substantive developments in the bio-oil commercialisation venture following these early leads. Table 6.1 summarises the major bio-oil production plants in Europe and the rest of the world and their status as on October 2020.

This recent commercialisation of bio-oil seems to have evolved along two different routes: companies that produce the bio-oil and those who use it for the production of district heating, electric power generation and other applications via CHP plants.

A major player in the production of bio-oil is **BTG Bioliquids** [11]. This commercial concern produces Fast Pyrolysis Bio-oil (FPBO) using rotating cone reactor technology at the following European sites:

- Empyro site at Hengelo, Netherlands which started in 2015;
- Green Fuel Nordic's Lieksa site in Finland, which commenced operations in 2020;
- Pyrocell, in Gavle, Sweden which was expected to start production in 2021 [11].

A corporate service provider that caters for district heating and electricity generation needs is **Savon Voima** of Finland [12].

Box 6.2 provides a brief overview of this Finnish facility.

Box 6.2. Case study of pyrolysis oil power CHP: The Savon Voima Plant at Joensuu, Finland.

Case Study: Bio-oil for the CHP power plant at Joensuu, Finland.

Brief background facts:

- A CHP plant existed at Joensuu, Finland before 2013. In November 2013, Fortum/Valmet commissioned a fast pyrolysis plant there. The plant began producing bio-oil from forest residues and other wood-derived biomass.
- On 20 December 2019, the Fortum/Valmet biorefinery at Joensuu was sold to Savon Voima.
- In December 2020, BTG Bioliquids' Green Fuel Nordic plant at Lieksa, Finland began supplying bio-oil to Savon Voima.

Savon Voima is currently Finland's sixth-largest network company, providing the transmission of electricity to 118 000 customers. Its facilities include 11 hydropower plants, and CHP plants at Lisalmi, Pieksamaki and Joensuu. Its current electrical and thermal power generation are 52 MW_e and 180 MW_{th} respectively [13].

(See research question 6.2—Chronology of commercialisation of bio-oil.).

6.5.4 Biogas-fired power plants

Biogas, which is a mixture of biomethane and carbon dioxide produced in biodigesters and landfill sites, can be used as fuel for thermal power generation plants. The technology for doing this is well-known and mature.

6.5.4.1 Biogas—global perspective

Biogas-fired power plants currently make a small but significant contribution to the global electrical power generation mix. According to IRENA [14], biogas power generation contributed 20 108 MW, or 0.72% to the global renewable energy based electrical power generation in 2020. As modern renewables such as biogas power generation formed 11.2% of the total final energy consumption in 2019, this amounts to approximately 0.08% of total global electricity generation in the same year.

While its contribution to the renewable energy (RE) power generation scene may be small, biogas still ranks favourably compared to several other RE power sources. Table 6.2 shows the relative standing of biogas to other sources of RE power generation today.

The table clearly reveals the relative preponderance of biogas over RE sources such as liquid biofuels, geothermal and all forms of marine energy in power generation.

According to the World Biogas Association [15], there are 50 million micro-digesters used in the domestic/farming context, 132 000 medium to large scale digesters supplying fuel for CHP plants, and some 700 plants used to upgrade biogas

Table 6.2. Global biogas-based power generation relative to other RE power generation sources and the total global final energy production in 2020 (data source: [14]).

Source	Installed capacity (MW) in 2020	Sub-total for sector (MW)	Share of global RE power generation (%)	Share of global final energy consumption (%)
Hydro (mixed plants)	57 388		2.05	0.23
Renewable hydro	1 154 313		41.29	4.62
Total hydro		**1 211 701**		
Wind (onshore)	698 043		24.97	2.80
Wind (offshore)	34 367		1.23	0.14
Total wind		732 410		
Solar	709 674		25.38	2.84
Total solar		709 674		
Bioenergy (solid biofuels)	89 042		3.19	0.36
Bioenergy (renewable municipal waste)	15 414		0.55	0.06
Bioenergy (Biogas)	***20 108***		***0.72***	***0.08***
Bioenergy (liquid biofuels)	2637		0.09	0.01
Total bioenergy		**127 201**		
Geothermal	14 013		0.50	0.06
Total geothermal		**14 013**		
Marine energy	527		0.02	0.002
Total marine energy		**527**		
Total RE power generation	**2 795 526**	**2 795 526**	**100.0%**	**11.2%**

to biomethane in the world today. Yet this accounts for merely 1.6–2.2% of the total global potential for producing biogas, revealing the vast potential for this source of energy that still remains untapped.

Examples of biogas power generation by country/region at the end of the last decade are given in table 6.3.

Table 6.3. Biogas power plant installed capacities by country/region (source: [15]).

Country/region	Number of CHP plants	Installed capacity
Europe	17 783	10.5 GW
Germany	9494	4.8 GW
China	6972	
USA	2200	977 MW
India		300 MW (estimated)
Canada	180	196 MW
World total		**20 108 MW**

6.5.4.2 Basic biochemistry of biogas production

Biochemical conversion of biomass into biogas occurs through the process known as **anaerobic digestion** and **methane fermentation**. As seen in chapter 4, the essential steps in this process, which are carried out by consortia of bacteria working in sequence, consist of hydrolysis, acidogenesis, acetogenesis and methanogenesis.

6.5.4.3 Technology requirements

The conventional techniques for the conversion of biogas energy to electricity is via a thermal power plant. This usually requires combustion of the biogas and the subsequent conversion of the heat energy. The appropriate technology is therefore either a gas turbine or reciprocating engine that can provide the prime mover for a thermal power plant.

6.5.4.4 Feedstock for biogas production

Feedstock is available from a large range of sources spanning the agricultural and industrial sectors as well as sources associated with municipalities [15]. They include but are not limited to

- agriculture—animal manure, crop residues, energy crops;
- industry—food industry (waste food, expired food, fats, oils); sugar industry, dairy industry, breweries and distilleries;
- municipal sources—MSW, sewage waste, garden waste.

Box 6.3 provides a case study of a biogas combined heat and power plant in Hamburg, Germany.

Box 6.3. The biogas power plant: Biowerk Hamburg GmbH & Co. KG at Stellinger Moor in Germany.

Situated at Stellinger Moore in Hamburg, Germany, the plant commenced operations in April 2006 [16, 17]. It is a joint venture partnership of three enterprises from the waste-to-energy industrial sector, comprising:

- Biocycling GmbH (47.5%—collection of expired food from supermarkets and leftover food from restaurants);
- ETH Umwelttechnik GmbH (5%—treatment of organic waste for utilisation in agriculture); and
- SRH Beteiliungs GmbH (47.5%—collection and disposal of household waste in Hamburg, operation of waste treatment plants).

The plant uses biogas produced from industrial, commercial and municipal wastes to produce heat for district heating and power generation. The process involves the fermentation of organic feedstock for the production of biogas, which is used for power generation and district heating via a CHP plant. The electricity is fed to the grid, while the heat generated is provided to a sports stadium for hot water and air-conditioning.

Table 6.4 provides further details of the production and output of the plant.

Table 6.4. Details of the production and output of the Biowerks Hamburg CHP at Stellinger Moor in Hamburg, Germany (data source: [16, 17]).

Production category	Quantity
Feedstock	
Expired foodstuff	15 000 tons yr^{-1}
Water	3450 tons yr^{-1}
Mixed waste	2000 tons yr^{-1}
Used oil and fat	2850 tons yr^{-1}
Biogas production	330 m^3 h^{-1}
Technology and output	
Engine	MWM Deutz 12 Zylinder V reciprocating engine
Electrical power produced	1.021 MW
Thermal power produced	1.070 MW
Overall efficiency	82.6%

Exercises and questions

Exercises

Exercise 6.1. Types of thermal power plants.

Determine the specific components (type of heat engine and balance of plant) of the thermal power plants that use the following energy sources:

(i) solid (woody) biomass used in an incineration power plant;
(ii) natural gas-fired power plant;
(iii) pyrolysis oil-fired power plant.

Exercise 6.2. The thermodynamics of a thermal power plant.

Examine the contents of figure 6.4 and correlate the physical features of a thermal power plant with the thermodynamic principles outlined in the previous chapter for a heat engine operating on the Rankine cycle. In particular,

a) What effect does condensing the spent steam from the turbine have on the efficiency of the power plant?
b) What is the purpose of reheating the steam from the HPT by sending it to the boiler?

(Hint: Relate the efficiency of the heat engine to the work done by it.)

Exercise 6.3. Efficiency of combined cycle power plants.

Consider a combined cycle power plant (CCPP) consisting of two steam turbines linked together in tandem, so that all the heat rejected to the cold reservoir by the first engine becomes the input heat (Q_h) for the second.

a) If the temperature at the input of the first engine is 600 °C and the temperature at the output is 400 °C, what is the Carnot efficiency of this engine? (Note: Do not forget to convert the temperature to Kelvins.)
b) If the second engine has a temperature of 40 °C at its output, calculate its Carnot efficiency, and use this to evaluate the combined efficiency of the whole system.

Questions

Question 6.1. Power output of an IGGPP.

An integrated gasification combined cycle power plant (IGCCPP) is to be built that uses fuelwood from an energy farm. The plant consists of a gasifier that feeds syngas to a gas turbine with an exhaust temperature of 360 °C. This gas is used to drive a steam engine which has an exhaust temperature of 40 °C. The electrical generators that are driven by the gas turbine and this engine both have efficiencies of 80%.

If the output power delivered by the gas turbine generator is 5 MW, calculate the output power of the steam turbine generator. Assume that the gasifier is 70% efficient, the gas turbine is 45% efficient, and the efficiency of the steam turbine is 60% the Carnot efficiency. (Hint: you can assume that all the heat lost by the gas turbine is used to drive the steam turbine.)

Research question 6.1. Plant-derived biomass and MSW as fuels for power plants.

Refer to research question 3.1 of chapter 3, which deals with pollution and other issues relating to power plants using fossil fuels and municipal solid wastes for incineration plants.

Carry out a similar study of power plants using plant-derived biomass. Compare the results with that for the waste-to-energy (WTE) plant of chapter 3 and make appropriate comments regarding the relative merits of the two methods of power generation. Tabulate your results under the headings: Calorific value of fuel, Pre-treatment of fuel, Emissions and Health Hazards, Preferred Technology.

After visiting the website of the Confederation of EU WTE Plants (www.cewep.eu) or following your own independent research, comment on the current status of WTE incineration plants in the world today.

Research question 6.2. Chronology of the commercialisation of bio-oil.

Provide a detailed chronology, starting early last decade, of the global commercialisation of bio-oil produced from the fast and catalytic pyrolysis of ligno-cellulosic feedstock.

The account should include the private enterprises, research institutions and government agencies involved in the development. It should also provide brief descriptions of the processes and technologies involved in both the production of the bio-oil and its use in heating and electricity generation.

References

[1] Mane S D 2016 Cogeneration in Indian sugar industry: a review. *Int. J. Sci. Eng. Appl. Sci.* **2** 30–40 [accessed 12 November 2022]. Available from www.ijseas.com

[2] Chandra V V 2020 A whole-of-system analysis of sugarcane bioenergy generation in Fiji *PhD Thesis* The University of the South Pacific p 147

[3] Raj N T, Iniyan S and Goic R 2011 A review of renewable energy-based co-generation technologies *Renew. Sustain. Energy Rev.* **15** 3640–8

[4] Ensinas A V, Nebra S A, Lozano M A and Serra L 2006 Analysis of cogeneration systems in sugar cane factories—alternatives of steam and combined cycle power plants *Proc. ECOS 2006 Aghia Pelagia, Crete, Greece. July 2006* 1177–84

[5] Fossil fuel power station. Wikipedia [accessed 13 November 2022]. Available from https://en.wikipedia.org/wiki/Fossil_fuel_power_station. Attribution: Tennessee Valley Authority.

[6] Birru E 2016 *Sugar cane industry overview and energy efficiency considerations. Report 01/2016* (Department of Energy Technology, KTH School of Industrial Engineering and Management, Stockholm) p 61

[7] Finding the right market for your FPBO. btgBioliquids [accessed 4 November 2022]. Available from https://btg-bioliquids.com/markets/

[8] Fortum's bio-oil plant commissioned in Joensuu. Valmet [accessed 14 November 2022]. Available from https://valmet.com/media/articles/all-articles/fortums-bio-oil-plant-commissioned-in-joensuu/

[9] Bio-oil via pyrolysis/thermochemical conversion. ETIP Bioenergy [accessed 14 November 2022]. Available from https://etipbioenergy.eu/value-chains/products-end-use/intermediates/bio-oil-tall-oil?highlight=WyJweXJvbHlzaXMiXQ

[10] Commercial plants. Pyrowiki [accessed 14 November 2022]. Available from http://pyrowiki.pyroknown.eu/index.php/Commercial_plants

[11] We replace fossil fuels. btgBioliquids [accessed 14 November 2022]. Available from https://btg-bioliquids.com/

[12] Eastern Finnish power for energy savings. Savon Voima [accessed 14 November 2022]. Available from https://savonvoima-fi.translate.goog/?_x_tr_sl=fi&_x_tr_tl=en&_x_tr_hl=en&_x_tr_pto=sc
[13] Bioenergy factsheet. ETIP Bioenergy [accessed 14 November 2022]. Available from https://etipbioenergy.eu/images/ETIP_B_Factsheet_Biomass_Fast_Pyrolysis_feb2020.pdf
[14] Statistics time series. IRENA [accessed 14 November 2022]. Available from https://irena.org/Data/View-data-by-topic/Capacity-and-Generation/Statistics-Time-Series
[15] Global potential of biogas. World Biogas Association. June 2019 [accessed 14 November 2022] Available from https://worldbiogasassociation.org/global-potential-of-biogas/
[16] Overview table for shining examples of biogas plants. Yumpu. Intelligent Energy. [accessed 1 May 2022]. Available from https://yumpu.com/en/document/read/34970638/overview-table-for-shining-examples-of-biogas-plants-a-biogas-
[17] Renewable energy from organic waste – the biogas plant Stellinger Moor in Hamburg. Biowerk Hamburg [accessed 1 May 2022]. Available from https://biowkk.eu/wp-content/uploads/2015/02/download-35.pdf

Chapter 7

Electrochemical conversion and storage of energy

The chemical energy stored in biomass can be extracted, stored and used as electrical energy directly via fuel cells. Batteries provide another means of storing energy in electrochemical form. Both these devices are gaining importance in the emissions reduction plans of many countries. This chapter describes the roles of electrolysers, fuel cells and batteries in electrochemical energy storage and use by first showing how the three devices are related to each other. It then surveys the types of fuel cells and batteries that are available today, and traces the evolution of the earliest (non-rechargeable) batteries to the latest rechargeable batteries. It ends with a brief look at current research in novel battery development.

7.1 Introduction

In chapter 4, it was pointed out that **fuel cells** were devices that converted the chemical energy of molecules directly into electrical energy. As this latter form of energy is in the highest demand amongst energy end users, it is of interest to see how such a technology relates to bioenergy utilisation.

A medium for **chemical energy storage** that is gaining increasing importance is hydrogen. The fuel cell of relevance is the **hydrogen fuel cell**. The device that converts electrical energy into the chemical energy stored in hydrogen is called an **electrolyser**.

Hydrogen itself can be classified as 'totally green' if its carbon footprint is zero. This is a very good approximation when it is produced via the **electrolysis of water** using 'zero' carbon footprint renewable energy as the electrical energy source.

A well-known form of energy storage is **electrochemical storage**. This is achieved through the use of **batteries** (more aptly called **electrochemical dry cells**). It is of interest to learn how these energy storage devices and media are related.

doi:10.1088/978-0-7503-3555-3ch7

In chapters 11 and 13, it will be seen how green hydrogen and battery storage play crucial roles in the reduction of carbon emissions, both as fuel sources for hydrogen fuel cell EVs (FCEVs) and as energy storage devices in grid energy storage. These energy storage media make significant contributions to the net-zero plans of many countries.

This chapter begins by firstly obtaining a clear picture of what fuel cells, electrolysers and batteries are, and how these devices are related to each other. This is facilitated by first introducing the electrochemical cell, which is the earliest electrochemical device used for the conversion of chemically-stored energy into electricity.

Section 7.2 provides a detailed introduction to the fuel cell, and gives brief descriptions of the seven types that are available today. Batteries are described in section 7.3. These are divided into **primary cells**, which are **non-rechargeable**, and **secondary cells** which are **rechargeable**. The development in the design and performance of the primary cells is traced from the most basic and earliest zinc–carbon batteries to the highest-performing (but most expensive) lithium batteries. The evolution of the rechargeable batteries from (the earliest) **lead-acid batteries** to the most recent **lithium-ion batteries** is then described. The last section takes a brief look at the research and development into new types of rechargeable battery design.

7.1.1 What are electrolysers, fuel cells and batteries?

To obtain a better insight into how fuel cells work, it will be helpful to first have an understanding of how fuel cells, electrolysers and batteries are related. One begins by noting that these three devices are involved in the conversion of chemical energy into electrical energy via **electrochemical processes**. This is quite different from the conversion process used by thermal power plants. These employ three other processes, starting from conversion of the **chemical energy** of biomass into **heat energy** through combustion, followed by the conversion of the heat energy into **mechanical energy** by heat engines, and finally into electricity via electrical generators.

Let us begin by elaborating on the roles of the three devices in energy transfer/conversion:

- **An electrolyser is a device that converts electrical energy into chemical energy stored in the molecules of a substance**—the most familiar example is the use of an electrolyser in the electrolysis of water, in which an electrical current induces the reaction

$$H_2O + \Delta E \rightarrow 2H_2 + O_2 \tag{7.1}$$

  Here, ΔE is the electrical energy used in the electrolysis, and the chemically-stored energy is the energy stored in the hydrogen molecules.
- **A hydrogen fuel cell is a device that converts the chemical energy stored in hydrogen to electrical energy**. The hydrogen energy source is usually unlimited (i.e. is available continuously).
- **A battery (i.e. an electrochemical cell) is a device that converts chemical energy stored inside it to electrical energy**. Its supply of energy is fixed to that contained in the chemicals stored inside the battery.

7.1.2 Introducing the electrochemical cell

To see how electrolysers, fuel cells and batteries are related, it is best to use the language and symbolism of the prototypical **electrochemical cell** (also called a Galvanic cell) [1]. This device utilises the tendency of atoms to attract or give up electrons to produce an electrical current in a circuit that can be usefully applied to operate electrical devices. The tendency of an atom to attract or give up electrons is called its **electronegativity** or **electropositivity**, respectively.

Metals have a greater tendency to give up electrons (i.e. are more electropositive) than non-metals. When a metal plate is placed in an aqueous solution containing its own ions, the atoms in the metal ionise to go into the solution. This arrangement of a metal plate placed in its own ionic solution is called a **half cell** [1, 2]. The ionising atom leaves behind the free electrons on the metal plate.

The tendency to ionise is determined by the (half cell) potential, measured in volts (V). Suppose a half cell of another metal is formed in the same manner and its metal plate is connected to the first through a metal wire. If an aqueous pathway (called a salt bridge) between the two cells is also provided, then:

- a current (i.e. a flow of electrons) will flow from one half cell to another via the metal conductor, and
- positive ions will flow between the two cells via the salt bridge to complete the circuit.

The assembly is called an **electrochemical cell**, and an example is shown in figure 7.1, which shows a copper half cell (called the anode) connected to a silver half cell (the cathode).

The two metal plates are called **electrodes**, and the ionic solutions form the **electrolyte**.

The size of the current in the external circuit will depend on the difference of the potentials (called the **potential difference**) between the two cells.

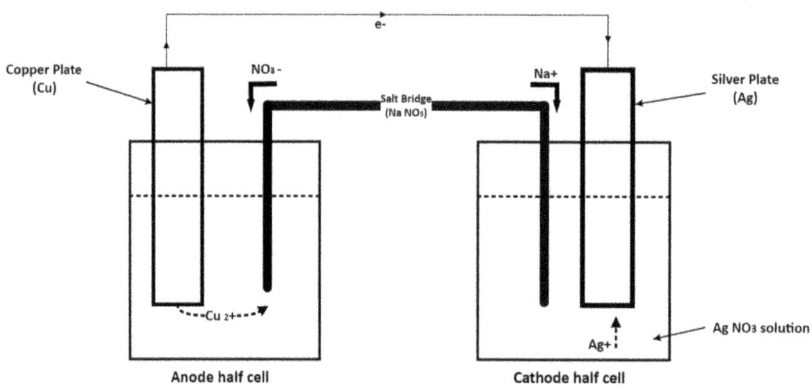

Figure 7.1. A schematic diagram of an electrochemical cell, showing half cells made from copper (the anode) and silver (the cathode). A current flows between the two electrodes (the copper and silver plates) when they are connected together by a metal wire and salt bridge as shown.

How this current flow originates may be explained in terms of the chemical reactions that occur between the electrodes and the solutions. The entire reaction process may be viewed as two halves of a **redox reaction**, where one metal (copper in the above case) is **oxidised**, and the ion of the other metal (silver) is **reduced**. Each half cell carries out half of this reaction. At the copper electrode (the anode) the half reaction is:

$$Cu \rightarrow Cu^{2+} + 2e^- \tag{7.2}$$

At the silver electrode (the cathode) the other half reaction is:

$$2Ag^+ + 2e^- \rightarrow 2Ag \tag{7.3}$$

and the resulting full redox reaction is:

$$Cu + 2Ag^+ \rightarrow Cu^{2+} + 2Ag \tag{7.4}$$

What is notable about this reaction is that it is **spontaneous** (i.e. the overall redox reaction will take place, and the current will start to flow, as soon as the wire is connected). The reaction will ideally keep taking place (and the current will keep flowing) until one of the electrodes dissolves completely in the solution.

What is the source of energy that drives this reaction? The most obvious answer is that it is the potential difference between the two electrodes. Its value in volts (V) can be measured directly using a voltmeter, and the size of the current (I) measured using an ammeter. The energy generated in the outer circuit per unit time (i.e. the power P) can then be calculated simply from the expression:

$$P = \mathrm{IV} \tag{7.5}$$

An equivalent answer is to say that the energy came from the chemical energy stored in the metal/electrolyte systems making up the two half cells. This cell provides the heart of the Leclanche dry cell (or battery, as it is called in common parlance). This is described in more detail later.

7.1.3 How are electrolysers, fuel cells and batteries related?

Returning to the definitions of the hydrogen fuel cell and the electrochemical cell given above, we see that **the hydrogen fuel cell is like a battery, the difference being that whereas the source of energy in the battery is the energy stored in the atoms of the metal electrodes, it is the hydrogen molecules in the case of the fuel cell**. As noted above, another important difference between the two is that the energy source in the fuel cell is made available continuously, while it is fixed in the case of the battery.

The full picture gained from the above analysis of electrolysers, fuel cells and batteries is summarised in box 7.1.

(See exercise 7.1—Electrolysers and electrochemical cells.)

Box 7.1. Comparing electrolysers, fuel cells and batteries.

A **battery** is a specific commercial implementation of the principle of the prototypical **electrochemical cell**. It produces electrical energy from a fixed amount of chemical energy stored inside it.

A **hydrogen fuel cell** is like a battery, but produces electricity from hydrogen. Unlike the battery, the energy source (hydrogen) of the fuel cell is made available continuously to the cell.

An **electrolyser** is a device that produces chemical substances (such as hydrogen) from electrical energy (i.e. it converts electrical energy to chemical energy). Its function is opposite to that of the fuel cell.

7.2 Types of fuel cells

7.2.1 Introduction

Fuel cells were first used to provide power for spacecraft from as early as the 1950s. Their use today has diversified to economic sector applications such as vehicle transportation and stationary power generation [3].

Figure 7.2 gives a schematic diagram of a hydrogen fuel cell.

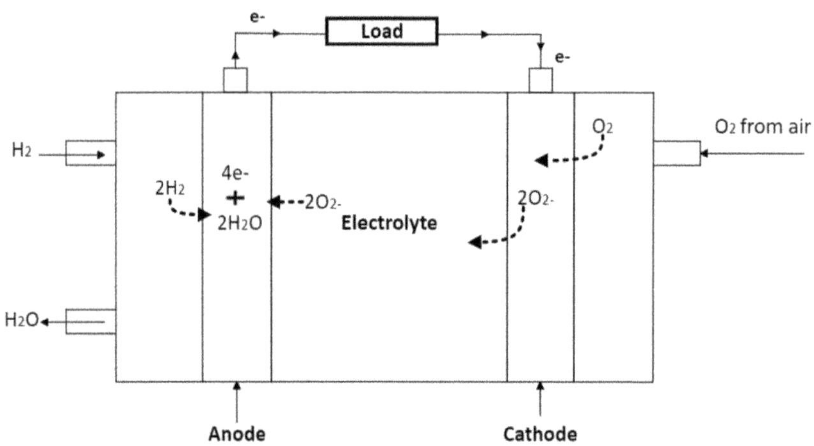

Figure 7.2. A schematic diagram of a hydrogen fuel cell.

The hydrogen fuel cell typically uses a supply of hydrogen, either from a direct source or through the reforming of natural gas, and oxygen derived from the atmosphere. The gases enter the device from opposite sides. Hydrogen enters at the anode and combines with O^{2-} ions to produce two molecules of water with the release of electrons. The oxygen ion is produced at the cathode when incoming oxygen combines with electrons arriving at the cathode from the anode via the external conductor. A catalyst is almost always involved in the processes.

The electrode reactions are:
Anode reaction

$$2H_2 + 2O^{2-} \rightarrow 2H_2O + 4e^- \tag{7.6}$$

Cathode reaction

$$O_2 + 4e^- \rightarrow 2O^{2-} \tag{7.7}$$

This yields the overall redox reaction by summation as:

$$2H_2 + O_2 \rightarrow 2H_2O. \tag{7.8}$$

7.2.2 Fuel cell types

Many different types of fuel cells exist [4]. They are generally categorised by the type of electrolyte they use into:

- liquid electrolyte fuel cells (where the electrolyte is an alkali, molten carbonate or phosphoric acid);
- solid electrolyte fuel cells (polymer electrolyte membrane (PEM), also called proton exchange membrane fuel cells);
- solid oxide fuel cells (where the electrolyte is the solid oxide of a ceramic material).

The various types of fuel cells that exist today or are being actively developed include:

- polymer electrolyte membrane (PEM) fuel cells;
- alkaline fuel cells (AFCs);
- direct methanol fuel cells (DMFCs);
- phosphoric acid fuel cells (PAFCs);
- molten carbonate fuel cells (MCFCs);
- solid oxide fuel cells (SOFCs);
- reversible fuel cells (RFCs).

A brief description of each follows.

7.2.2.1 Polymer electrolyte (PEM) fuel cells

Polymer electrolyte membrane (PEM) fuel cells (also known as proton exchange membrane fuel cells) employ solid polymer membranes as electrolytes, and a platinum or platinum alloy catalyst which is applied to their porous carbon electrodes. They obtain hydrogen from either fuel tanks or reformers (which extract hydrogen from natural gas, biogas or coal-derived gases), and oxygen from the air. They typically have low weight and smaller volumes, making them easier to handle.

Operating temperatures are relatively low (around 80 °C). However, the platinum catalyst used to strip the electrons from hydrogen is expensive, contributing significantly to system costs. The catalyst is also highly sensitive to carbon monoxide poisoning. PEM fuel cells are appropriate for use in vehicles such as cars, buses and trucks.

7.2.2.2 Alkaline fuel cells (AFCs)

Alkaline fuel cells (AFCs) are liquid electrolyte fuel cells that use aqueous potassium hydroxide as the electrolyte and relatively inexpensive metals as the catalyst at the anode and cathode. They are closely related to the PEM fuel cell, with the difference that they use an alkali instead of an acid membrane for the electrolyte. They have high performance and efficiencies, sometimes exceeding 60%.

A drawback of AFCs is their susceptibility to poisoning by carbon dioxide. These liquid electrolyte systems also suffer from other issues such as wettability, increased corrosion and difficulties handling differential pressures.

7.2.2.3 Direct methanol fuel cells (DMFCs)

Unlike most fuel cells, which use hydrogen as the energy source, direct methanol fuel cells (DMFCs) are powered by pure methanol, which is mixed with water and fed directly to the anode of the fuel cell. Because of the favourable fuel density, direct methanol fuel cells do not have the fuel storage issues of other types of fuel cells. As methanol is a liquid, it is easier to transport and distribute. Amongst the common uses of DMFCs are applications such as cell phones or laptop computers.

7.2.2.4 Phosphoric acid fuel cells (PAFCs)

Phosphoric acid fuel cells (PAFCs) employ liquid phosphoric acid as the electrolyte, porous carbon as the electrodes and platinum as the catalyst. PAFCs are considered to be a mature fuel cell technology, and were amongst the first to be commercialised. They are used for stationary power generation, and in some cases provide the energy source for larger vehicles such as buses.

PAFCs are more tolerant than PEM cells to impurities in fossil fuels that have been reformed into hydrogen than PEM cells. Their electricity-generating efficiency is low, but they can operate with up to 85% efficiency when used in the combined heat and power (CHP) mode. PAFCs also have less power densities (i.e. power produced per unit mass) making them heavy. They use platinum as the catalyst, which adds to their cost.

7.2.2.5 Molten carbonate fuel cells (MCFCs)

Molten carbonate fuel cells (MCFCs) are high-temperature devices that use molten carbonate salt suspension in a porous ceramic lithium aluminium oxide matrix as electrolytes. They operate at temperatures of around 650 °C, and use non-precious metals as catalysts, contributing to a reduction in costs. MCFCs can reach high efficiencies (up to 65%) when used for electricity generation alone, and up to 85% in the CHP mode.

A notable feature of MCFCs is that, as a result of their high-temperature operation, they do not require an external reformer to reform fuels such as natural gas and biogas. The high temperature of operation facilitates thermal reforming of methane and other hydrocarbons to hydrogen within the fuel cell through the process of internal reforming. The main disadvantage of these devices is their short lifetimes due to their harsh operating conditions. The high operating temperatures and the corrosive nature of the electrolyte makes the current generation MCFCs highly susceptible to wear and tear and prone to failures.

7.2.2.6 Solid oxide fuel cells (SOFCs)

Solid oxide fuel cells (SOFCs) employ ceramics (based on perovskites) as their electrolyte. Their electricity generation efficiency is as high as 60%, and when operated as combined heat and power plants (CHPs) their overall efficiencies may reach or exceed 85%.

These fuel cells need very high temperatures (up to 1000 °C) for their operation. This facilitates internal reforming of fuels, making way for the use of a range of fuels. This allows the use of natural gas, biogas and coal-derived gases as fuels.

The high-temperature operation, however, results in long startup times, and requires thermal shielding for heat retention and personal protection. The high operating temperatures also place stringent durability requirements on materials. Lower-temperature SOFCs (those capable of operating below 700 °C) are now being investigated.

7.2.2.7 Reversible fuel cells

Reversible fuel cells are devices that can act both as fuel cells and electrolysers. When operated in normal mode, they generate electricity from hydrogen and oxygen. When operated in reverse mode (i.e. when an electrical potential is applied to them to drive a reverse current) they produce hydrogen and oxygen from water. As they can employ renewable energy sources such as wind, solar or hydro, they are capable of producing emission-free hydrogen (i.e. green hydrogen) when used in electrolysis mode.

Such devices are appropriate for energy supply systems that generate and store excess clean energy using hydrogen storage. They are particularly important in the efficient utilisation of intermittent renewable energy resources such as wind and solar energy.

7.2.3 Summary of essential properties of fuel cells

The above list shows that there is a wide variety of fuel cells possible, each type having its advantages and disadvantages depending on its particular properties. Perhaps the most important of these properties is the efficiency of the fuel cell. This includes:

- for fuel cells dedicated to power generation alone, the efficiency of power generation;
- for fuel cells that can be used as combined heat and power (CHP) sources, the combined power generation efficiency and thermal efficiency.

In summary, the most important criteria that decide the quality of fuel cells are:

- efficiency (for both stand-alone power generation and combined heat and power operation), which must be high;
- power density (the power generated per unit mass (or volume) of the fuel cell), which must be high;
- operating temperature, which must not exceed room temperature by any considerable amount;
- use of catalysts, which must not be too expensive;

- nature of the electrolyte, which must, amongst other things, satisfy health and safety requirements;
- susceptibility of electrodes to poisoning, which must be low or absent;
- ability to reform fuels into hydrogen, which adds versatility to the device;
- overall system costs, which must be low;
- system lifetime, which must be as long as possible.

(See question 7.1—Fuel cell properties and their required values.)

(See research question 7.1—State of the art of fuel cells.)

Fuel cells will be discussed again in chapters 11 and 13, in the context of using hydrogen as a fuel for transportation and power generation, and the potential role of hydrogen fuel cells in climate change mitigation technology. They will be considered as one of the possible technologies contributing to Australia's net-zero plan for emissions reduction by 2050.

7.3 Types of batteries

In section 7.1.1, a battery was described as a device that converts (stored) chemical energy to electrical energy, and it was pointed out that the electrochemical cell served as the prototype for all batteries in common use.

Commercially available batteries may also be pictured as (portable) self-contained power sources that produce electrical power as the need arises from chemicals stored inside them. Although the principle of operation of all batteries is essentially the same as that of the electrochemical cell, they occur in several variations depending on the choice of electrolytes and electrodes. They also come in a large range of physical forms and sizes. These can be divided into two broad categories: **primary batteries** and **secondary (rechargeable) batteries**.

Primary batteries are batteries that **cannot be recharged**. These are **disposable batteries** that need to be replaced once they have become fully discharged. Secondary batteries are those that can be recharged, simply by passing a current through them in a direction opposite to the current they produce when they are in operation.

7.3.1 Primary batteries

The three main types of primary (i.e. non-rechargeable) batteries are **zinc–carbon**, **alkaline**, and **lithium** batteries (also called **lithium-metal batteries**) to distinguish them from lithium-ion batteries, which are rechargeable [5].

7.3.1.1 Zinc–carbon batteries

These are the cheapest disposable dry cell batteries in everyday use for applications such as flashlights, transistor radios and other portable appliances, and are usually encountered as AA, AAA, C or D type dry cells. Invented in 1866 by Georges Leclanche, the zinc–carbon battery was the first dry cell battery to be introduced to the market.

The positive electrode consists of a mixture of manganese dioxide and powdered carbon surrounding a carbon rod, while the negative electrode is provided by zinc metal which forms the outer casing of the cell. The electrolyte is either an ammonium chloride or zinc chloride paste.

The electrode reactions are:

(Using ammonium chloride paste as electrolyte)

Anode (negative electrode): Oxidation reaction in which Zinc goes into solution as Zn^{2+} ions

$$Zn + 2Cl^- \rightarrow ZnCl_2 + 2e- \tag{7.9}$$

Cathode (positive electrode): Reduction reaction in which the manganese ion is reduced from the (IV) oxidation state to the (III) state (i.e. MnO_2 is transformed to Mn_2O_3) and an aqueous solution of ammonia is formed.

$$2\,MnO_2 + 2\,NH_4Cl + H_2O + 2e^- \rightarrow Mn_2O_3 + 2\,NH_4OH + 2\,Cl^- \tag{7.10}$$

Yielding the overall reaction:

$$Zn + 2\,MnO_2 + 2\,NH_4Cl + H_2O \rightarrow ZnCl_2 + Mn_2O_3 + 2\,NH_4OH \tag{7.11}$$

With $ZnCl_2$ as the electrolyte, the cathode reaction is:

$$2\,MnO_2 + ZnCl_2 + H_2O + 2\,e^- \rightarrow Mn_2O_3 + Zn(OH)_2 + 2\,Cl^- \tag{7.12}$$

Zinc–carbon batteries have a short self-life. The ammonium chloride reacts with the zinc causing the battery to 'leak' once the zinc metal casing is corroded.

7.3.1.2 Alkaline batteries

Alkaline batteries have the same voltage (of 1.5 V) at their terminals as the zinc–carbon (Zn–C) batteries, but have a higher energy density (i.e. energy per unit mass or volume) and last much longer. These batteries are similar to Zn–C batteries in their basic construction except that they replace the (acidic) electrolyte with potassium hydroxide (KOH), which is an alkali.

The half reactions are:

Negative electrode:

$$Zn + 2OH^- \rightarrow ZnO + H_2O + 2e^- \tag{7.13}$$

Positive electrode:

$$2MnO_2 + H_2O + 2e^- \rightarrow Mn_2O_3 + 2OH^- \tag{7.14}$$

Overall reaction:

$$Zn + 2MnO_2 \rightleftharpoons ZnO + Mn_2O_3 \tag{7.15}$$

These batteries have a much longer shelf-life than zinc–carbon (Zn–C) batteries, and are very popular, though more expensive. They are used in a range of household devices such as toys, flashlights, digital equipment including CD and MP3 players.

7.3.1.3 Lithium batteries

Lithium batteries (not to be confused with lithium-ion batteries) are high performance primary batteries with high energy densities (enabling smaller, light-weight packaging) and a very long shelf-life (far exceeding those of alkaline batteries). The basic design uses metallic lithium as the anode with a range of solid inorganic, organic or organo-metallic cathodes and electrolytes [6, 7]. The most common type consists of a cell using:

- metallic lithium as the anode;
- manganese dioxide as the cathode; and
- a salt of Li dissolved in organic solvent as the electrolyte.

During cell operation, lithium oxidises, producing Li^+ ions and electrons at the anode. The Li^+ diffuses through the electrolyte to the cathode while the electron reaches there via an external circuit to complete the redox reaction.

A particular example of a lithium primary battery is the **lithium–iodine (Li–I) battery**. This has metallic lithium as the anode, a solid complex of iodine in the form of the organic solid **poly-2-vinyl pyridine (P2VP)** as the cathode, and a solid **lithium iodide layer as the electrolyte** [7].

The two main issues with lithium primary batteries are safety concerns due to the reactivity of lithium and the high costs.

Lithium is one of the alkali metals, and is highly electropositive. This means that it gives up its electron (i.e. ionises) readily, and is therefore highly reactive. In particular, it **reacts violently with water**, and needs to be completely sealed to exclude any contact with moisture.

Secondly, lithium is a very expensive metal due to its scarcity as a mineral resource on Earth. This makes lithium batteries much more expensive than alkaline or Zn–C batteries.

7.3.1.4 Button/coin batteries

Over the last four decades, numerous forms (or packagings) of dry cell batteries have begun to emerge on the consumer market. One physical type that deserves mention is the so called **button cell** or **coin cell**. These now find wide usage in all types of electronic devices, and have been used in quartz watches and other portable electronic devices since the 1980s. These various forms of batteries are now identified under a universal system of classification established by the International Electrochemical Commission (IEC) [8, 9].

Many of the batteries available today are merely special packagings of the basic alkaline or lithium primary batteries discussed above, often consisting of a series of individual coin cells packed into cylindrical forms and other formats. The central part of the casing of the basic cell is the negative electrode, and is made of either metallic lithium or zinc. The outer casing is the positive electrode, made of a metal oxide such as manganese, silver or copper oxide. Two particular examples of these cells are the CR2032 lithium battery, and the LR44 alkaline battery (see [9] for an explanation of the nomenclature). Figure 7.3 exemplifies the range of such batteries that are now available commercially.

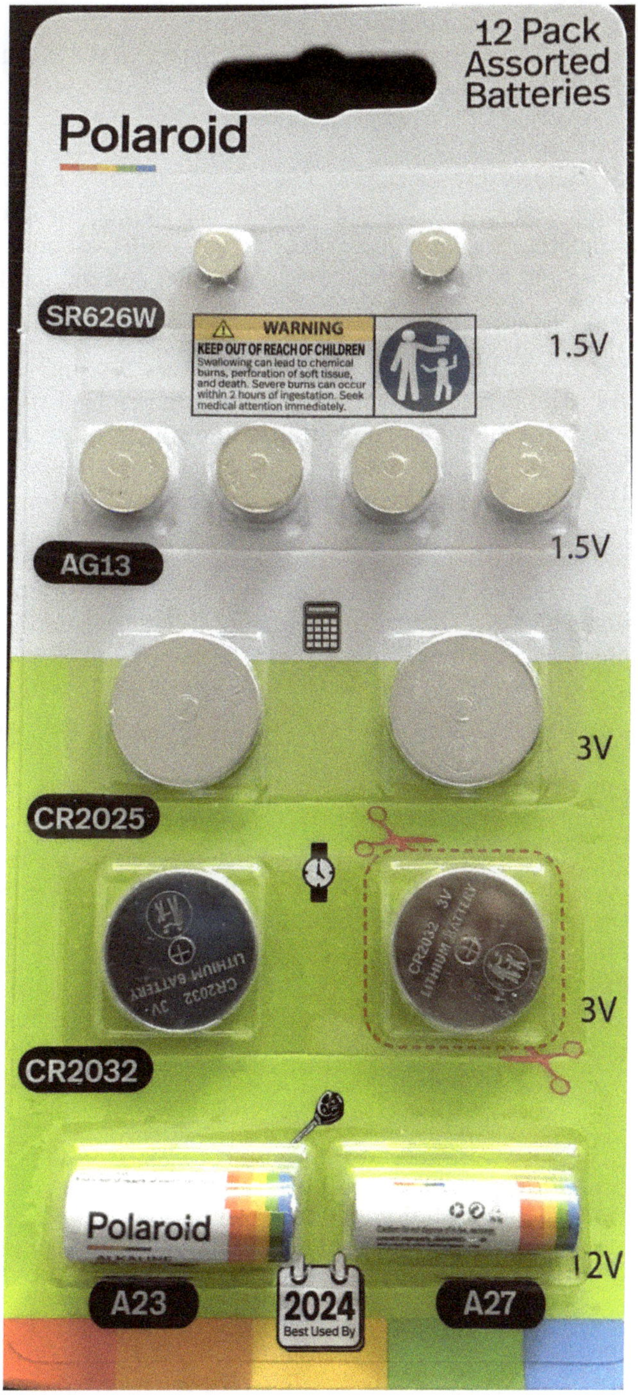

Figure 7.3. Examples of button/coin cells in use in a diverse range of portable electronic devices today.

7.3.1.5 Primary batteries in application

In summary, it is instructive to compare the performance and use of the three types of primary batteries available commercially today.

7.3.1.5.1 Zinc–carbon batteries

These are used as a cheap, disposable power supply (in physical formats such AA and AAA, C and D batteries) for low-cost applications such as flashlights, toys, radios and other electronic devices. Compared to other types, Zn–C batteries have been in existence the longest, but have the lowest energy densities and the shortest shelf-lives.

7.3.1.5.2 Alkaline batteries

Alkaline batteries provide longer-lasting power sources for digital equipment, in most cases where regular replacement of batteries is possible. Examples include a wide variety of portable digital equipment including watches, cameras, calculators, car keys, etc. where long battery life is an advantage. They are used in all the packagings available to the Zn–C batteries as well as in the button cell format.

7.3.1.5.3 Lithium primary batteries

Lithium batteries are used where longer-lasting batteries are required than that offered by alkaline batteries. They are especially useful in situations where long battery life is a necessary feature. This applies to situations where the battery needs to operate for extended periods (years) of time without need for replacement. An important example is their use in medical implants such as **heart pacemakers** and **defibrillators**.

7.3.2 Secondary (rechargeable) batteries

More commonly known as **rechargeable batteries**, these have become very popular since the arrival of portable digital information-related technology such as the mobile phone, the laptop, as well as digital cameras and MP3 players. More recently, rechargeable batteries have become indispensable as energy sources for electric vehicles (EVs). The four main types of rechargeable batteries consist of the **lead-acid battery**, **nickel–cadmium** battery, the **nickel–metal hydride (NiMH) battery** and the **lithium-ion** battery. Of these, the lithium-ion battery is currently enjoying the greatest popularity in both grid energy storage and EVs.

7.3.2.1 Lead-acid batteries

The most established form of the rechargeables is the lead-acid battery, which predates the digital era by at least a century. These batteries are still the work horse for vehicle electrical power requirements [10]. Invented in the 19th century, the lead-acid battery is typically made of several cells (each capable of providing a maximum potential difference of 2.1 V) connected together in series. Inside the cell, the two electrodes consist of lead plates. In the fully charged state, the positive is coated with a paste of lead dioxide and the negative consists of lead metal.

When fully discharged, both electrodes consist of lead sulphate coated on metallic lead plates, each electrode being in contact with the dilute sulphuric acid electrolyte.

In the **discharge process**, the half reaction at the negative electrode is:

$$Pb + HSO_4^- \rightarrow PbSO_4 + H^+ + 2e^- \quad (7.16)$$

The lead electrode acquires the two electrons and becomes negatively charged. This creates an electron flow from the negative to the positive electrode when connected externally by a conductor.

At the positive electrode:

$$PbO_2 + HSO_4^- + 3H^+ + 2e^- \rightarrow PbSO_4 + 2H_2O \quad (7.17)$$

Yielding the overall reaction:

$$Pb + PbO_2 + 2H_2SO_4 \rightarrow 2PbSO_4 + 2H_2O \quad (7.18)$$

The energy that drives the current is made available when the H^+ ions react with the O^- ions at the positive electrode to form water via the reaction:

$$4H^+ + 2\,O_2^- \rightarrow 2H_2O + \Delta E \quad (7.19)$$

where ΔE is approximately −880 kJ per 18 g of water. For a more detailed explanation of the chemistry, see [11].

Overcharging:

When high charging voltages are applied to the terminals, the battery acts as an electrolyser and produces hydrogen and oxygen gases from the water present in the dilute sulphuric acid electrolyte. This causes a safety hazard, especially in the case of sealed batteries, which necessitates appropriate safety measures [12].

7.3.2.2 Nickel–cadmium batteries

The original wet cell version of the nickel–cadmium (pronounced 'Nicad') battery was invented in 1899 [13]. It has a maximum terminal voltage (i.e. a total electromotive force) of 1.3 V, of which about 1.2 V is available at the terminals in normal operation. The cell consists of a positive electrode of nickel oxide hydroxide (NiO(OH)), a negative electrode of cadmium (Cd), and an alkaline electrolyte of potassium hydroxide (KOH).

The battery is available in a wide variety of packagings, ranging from portable sealed batteries that can be exchanged with Zn–C batteries (sizes AA, AAA and D), to large vented cells for both stand-by power supply needs and for transportation.

Reactions:

The **discharge process** comprises the following electrode reactions:

Positive electrode:

$$2NiO(OH) + 2H_2O + 2e^- \rightarrow 2Ni(OH)_2 + 2OH^- \quad (7.20)$$

Negative electrode:

$$Cd + OH^- \rightarrow Cd(OH)_2 + 2e^- \quad (7.21)$$

Overall reaction:

$$2NiO(OH) + Cd + 2H_2O \rightarrow 2Ni(OH)_2 + Cd(OH)_2 \quad (7.22)$$

The **recharge process**:

To recharge the battery, a potential is applied to the terminals to simply reverse the direction of current flow. This in turn forces the reverse of the above reactions to take place at the terminals.

Advantages:

The nickel–cadmium rechargeable battery has a much **higher energy density** than the lead-acid battery. It also has a very **large discharge rate**, enabling it to deliver large currents over short periods of time (<1 h).

Disadvantages:

Its main disadvantages are its higher costs (as compared to the lead-acid battery) and the **toxicity of cadmium**, which creates an environmental issue.

Uses:

Sealed NiCad cells are used for a range of portable electronic devices, including flashlights and power tools. Battery packs delivering ~7.2 V made from individual cells of 1.2 V are available for higher voltage applications. Large vented wet cells are used to power aviation and trains in addition to providing stand-by power supplies.

NiCad is now superseded by nickel–metal hydride batteries (due largely to the environmental concerns of cadmium and the superior performance of the latter batteries).

7.3.2.3 Nickel–metal hydride (NiMH) batteries

The nickel–metal hydride (NiMH) battery has effectively replaced NiCad batteries due mostly to the toxicity of cadmium. It has the same positive electrode as NiCad, but the negative electrode is replaced by a hydrogen-absorbing metal alloy (provided by an intermetallic compound). The NiMH battery has the same terminal voltage as the NiCad, and is used as a replacement for similar size (i.e. similarly packaged) non-rechargeable alkaline batteries. It has a slightly lower voltage but is less prone to explosions [14, 15].

The chemical reactions are as follows:

Positive electrode:

$$2NiO(OH) + 2H_2O + 2e - \rightarrow 2Ni(OH)_2 + 2OH^- \quad (7.23)$$

Negative electrode:

$$H_2O + M + e^- \rightarrow OH^- + MH \quad (7.24)$$

where M is an intermetallic compound.

NiMH batteries have 2–3 times the capacity of NiCad batteries and significantly higher densities. However, these features remain lower than those for lithium-ion batteries.

7.3.2.4 Lithium-ion batteries

The highly-popular **lithium-ion rechargeable battery** has a positive electrode made of a lithium cobalt oxide and a negative electrode made of lithium intercalated between layers of graphite.

During charging, the positive electrode loses Li^+ ions, which move through the electrolyte to the negative electrode. The electrons lost by the lithium atoms at the positive electrode flow through an external circuit to the negative electrode, where they combine with the arriving Li^+ ions. The reverse process occurs during discharge [16, 17].

The electrode reactions are:

Anode reaction:

$$LiCoO_2 \rightleftharpoons Li_{1-x}CoO_2 + xLi_+ + xe- \tag{7.25}$$

Cathode reaction:

$$xLi^+ + xe^- + xC_6 \rightleftharpoons xLiC_6 \tag{7.26}$$

where $x \sim 0.5$.

Overall reaction:

$$LiCoO_2 + xC_6 \rightleftharpoons Li_{1-x}CoO_2 + xLiC_6 \tag{7.27}$$

During charging, the reactions proceed from left to right and in the reverse direction during discharge.

(See exercise 7.2—Comparing EVs and gasoline engine cars.)

7.4 Battery research and development

Lithium-ion rechargeable batteries have one major drawback. It is the availability of the element lithium for battery manufacture. Lithium is one of several critical metals needed for global battery supply needs. In 2021, the world produced only 104 000 metric tons of this material [18]. The scarcity of this raw material has therefore been generating strong interest in the hunt for cheaper alternatives.

One such alternative is the **sodium-ion battery**. Because of the global availability of sodium (it is a constituent of table salt, a material found in abundance in our seas), this alkali metal is a much cheaper alternative for battery manufacture than lithium.

The basic principle of operation of the sodium-ion rechargeable battery is essentially the same as that of the lithium-ion battery [19]. The first sodium-ion battery cell was introduced by Faradion in 2015. This company was later acquired by the Indian-based Reliance Industries, which is now competing with the world's largest battery manufacturer, the China-based Contemporary Amperex Technology Company Limited (CATL) in the production of a market-ready product [20].

Currently, another area of active battery research is the production of electricity using **hydro-voltaic technologies**. These employ moisture gradients set up in 2D devices known as **Self-Sustained Electricity Generators (SSEGs)** or **Moisture Electric**

Generators (MEGs) to create electric potential differences between the moist and dry sides of the device.

The principle of operation of SSEGs is explained by Tan *et al* in their recent *Nature Communications* paper [21]. The device used by this group consists of two layers of cellulon paper. The moist-air facing layer is loaded with lithium chloride (LiCl) and is hygroscopic (i.e. attracts moisture) while the dry-air facing side is loaded with carbon black, and is hydrophobic, promoting the evaporation of the water. Voltages and currents of 0.78 V and 7.5 μA, sustainable for up to 10 days have been reported.

Recently, the University of New South Wales (UNSW) Nanoionic Materials Group, led by Prof. Dewei Chu, have developed MEGs based on graphene oxide, a two-dimensional nano-material that gives improved results. In their publications in the journals *Nano Energy* and *Applied Nano-Materials* [22, 23], they report the fabrication of a MEG using acidified graphene oxide films incorporated with small amounts of polyvinyl alcohol (PVA) that yields a voltage of 0.85 V and very high current densities of 92.8 μA cm^{-2} at a relative humidity of 80%. Their MEG shows voltage retention of over 2 h.

According to the *Nano Energy* paper, a MEG with such high voltage and current outputs could directly power such electronic devices as memristors and various sensors. By connecting up to four such devices in parallel/series, outputs as high as 38.95 μA and 3.38 V could be obtained.

(See research question 7.2—Understanding hydro-voltaic technology.)

Exercises and questions

Exercises

Exercise 7.1. Electrolysers and electrochemical cells.

a) What would the electrochemical cell in figure 7.1 become if a potential difference (i.e. voltage) was applied to its electrodes from an external source that drove the electron flow in the opposite direction?
b) Show the equations for the half reactions that take place, and sum them up to arrive at the resultant redox reaction.

Exercise 7.2. Comparing EVs and gasoline engine cars.

Using available data, 'How lithium-ion batteries work', available at [17] or otherwise,

a) Compare the range of an EV running on a lithium-ion battery to that of a gasoline (combustion) engine car.
b) Evaluate the range per unit energy (km MJ^{-1}) of the two technologies and comment.
c) Compare the other performance, environmental and economic parameters (including maximum speed, time for fuel recharge, availability of fuelling stations and price of the two cars) using a matrix, and comment.

Questions

Question 7.1. Fuel cell properties and their required values.

a) From the information available in the above overview of fuel cells or otherwise, construct a matrix of the criteria that needs to be fulfilled by an acceptable fuel cell, together with the acceptable value for each criterion. Use the list of criteria given in section 6.3.2.

b) Use the results of part (a) to determine the most appropriate type of fuel cell for:

   (i) a mobile phone;
   (ii) an electric bus;
   (iii) a grid-connected hydrogen energy storage system.

Research question 7.1. State of the art of fuel cells.

Do a literature search and

a) Report on the current status of development of fuel cells for commercial, industrial and power sector applications. In particular,

   (i) report on the minimum operating temperatures that have been attained for solid oxide fuel cells (SOFCs);
   (ii) comment on the other parameters that determine the quality of SOFCs.

b) Write a brief report on the current status of reversible fuel cells. In particular, discuss their use in grid-connected energy storage systems based on hydrogen.

Research question 7.2. Understanding hydro-voltaic technology.

The hydro-voltaic technologies discussed in section 7.4 are unique in that they appear to generate electricity from the presence of atmospheric moisture alone. It is therefore instructive to investigate the energetics of these devices.

a) Describe the SSEG fabricated by Tan *et al* (*Nature Communications* paper) and the MEG reported by Prof. Chu's group (*Nano Energy* paper) and outline the principle of operation of each. How do these two devices differ?
b) What is the necessary requirement for all such devices to work?
c) What is the energy source that produces the electrical energy in such devices?
d) What (if any) are the limitations of such devices in application?

References

[1] Electrochemical cells. LibreTexts Chemistry [accessed 20 November 2022]. Available from https://chem.libretexts.org/Bookshelves/General_Chemistry/Map%3A_Principles_of_Modern_Chemistry_(Oxtoby_et_al.)/Unit_4%3A_Equilibrium_in_Chemical_Reactions/17%3A_Electrochemistry/17.1%3A_Electrochemical_Cells

[2] The Cell Potential. LibreTexts Chemistry [accessed 20 November 2022]. Available from https://chem.libretexts.org/Bookshelves/Analytical_Chemistry/Supplemental_Modules_(Analytical_Chemistry)/Electrochemistry/Voltaic_Cells/The_Cell_Potential

[3] Combined Heat and Power Factsheet Series. UD Department of Energy, EE and RE. 2016 [accessed 20 November 2022]. Available from https://energy.gov/sites/default/files/2016/09/f33/CHP-Fuel%20Cell.pdf
[4] Types of Fuel Cells. Energy.gov. EE and RE [accessed 20 November 2022]. Available from https://energy.gov/eere/fuelcells/types-fuel-cells
[5] Batteries. Explain that stuff [accessed 20 November 2022]. Available from https://explainthatstuff.com/batteries.html
[6] Batteries. LibreTexts Chemistry [accessed 20 November 2022]. Available from https://chem.libretexts.org/Bookshelves/General_Chemistry/Map%3A_Chemistry_(Zumdahl_and_Decoste)/11%3A_Electrochemistry/11.5%3A_Batteries
[7] Lithium battery. Wikipedia [accessed 20 November 2022]. Available from https://en.wikipedia.org/wiki/Lithium_battery
[8] The nomenclature of the battery. Blue light [accessed 20 November 2022]. Available from http://e-bluelight.com/article-417-The+nomenclature+of+the+battery.html#:~:text=Since%201992%2C%20International%20standard%20IEC%2060086%20defines%20an,are%20batteries%20coded%20R20%2C%204LR25-2%2C%20CR17345%2C%20and%206F22
[9] Battery Nomenclature. Wikipedia [accessed 20 November 2022]. Available from https://en.wikipedia.org/wiki/Battery_nomenclature
[10] Lead-acid battery. Wikipedia [accessed 21 November 2022]. Available from https://en.wikipedia.org/wiki/Lead%E2%80%93acid_battery
[11] Schmidt-Rohr K 2018 How batteries store and release energy: explaining basic electrochemistry *J. Chem. Educ.* **95** 1801–10
[12] Hydrogen gas risk in battery charging rooms. Safe Environments [accessed 21 November 2022], Available from https://safeenvironments.com.au/hydrogen-gas/
[13] Nickel–Cadmium battery. Wikipedia [accessed 21 November 2022]. Available from https://en.wikipedia.org/wiki/Nickel%E2%80%93cadmium_battery
[14] Nickel–Metal hydride battery. Wikipedia [accessed 21 November 2022]. Available from https://en.wikipedia.org/wiki/Nickel%E2%80%93metal_hydride_battery#cite_note-7
[15] Batteries and Fuel Cells. LibreTexts Chemistry [accessed 21 November 2022]. Available from https://chem.libretexts.org/Bookshelves/General_Chemistry/Map%3A_Chemistry_-_The_Central_Science_(Brown_et_al.)/20%3A_Electrochemistry/20.07%3A_Batteries_and_Fuel_Cells
[16] Batteries and Fuel Cells. LibreTexts.Chemistry [accessed 21 November 2022]. Available from https://chem.libretexts.org/Bookshelves/General_Chemistry/Chemistry_(OpenSTAX)/17%3A_Electrochemistry/17.5%3A_Batteries_and_Fuel_Cells
[17] Lithium ion batteries. Explainthatstuff [accessed 21 November 2022]. Available from https://explainthatstuff.com/how-lithium-ion-batteries-work.html
[18] Major countries in worldwide lithium mine production in 2021. Statistica [accessed 19 November 2022]. Available from https://statista.com/statistics/268789/countries-with-the-largest-production-output-of-lithium/
[19] Sodium ion battery breakthrough. YouTube [accessed 19 November 2022]. Available from https://youtube.com/watch?v=cHNELRnJ_4Y
[20] Energy and Power/Sodium-ion battery market. Fortune Business Insights [accessed 19 November 2022]. Available from https://fortunebusinessinsights.com/industry-reports/sodium-ion-battery-market-101301

[21] Tan J, Fang S and Zang Z *et al* 2022 Self sustained electricity generator driven by the compatible integration of ambient moisture absorption and evaporation *Nat. Commun.* **13** 3643
[22] Zhu R, Zhu Y and Chen F *et al* 2022 Boosting moisture induced electricity generation from graphene oxide through engineering oxygen-based functional groups *Nano Energy* **94** 106942
[23] Feng Z, Hu G and Zhu R *et al* 2022 Two-dimensional nanomaterials for moisture-electric generators: a review *ACS Appl. Nano Mater.* **5** 12224–44

Part III

Liquid and gaseous biofuels for transportation and power generation

IOP Publishing

Chapter 8

Introduction to liquid biofuels

Liquid biofuels provide an alternative to the fossil-derived fuels used in transportation. This chapter discusses the use of these alternative fuels in land transportation by firstly introducing the nature and properties of such fuels, and their merits as agents of emissions reduction. The fuel standards and legislations used globally for diesel and petrol (gasoline) engines are next elaborated on in some detail, as the alternative fuels will need to comply with these standards if used in unmodified engines. Biofuels differ from the purely hydrocarbon-based gasoline and diesel fuels in certain ways, and thus require standards of their own. These are briefly discussed. The chapter ends with a case study of how Australia legislates its standards for transportation fuels.

8.1 Introduction

Part I of the book considered the role of biomass as an energy source. It began by surveying the forms of solid biomass available as sources of renewable energy, examined the properties of biomass that determined their energy content, and ended by detailing the methods commonly used to convert solid biomass from its original form to forms that were more amenable to use in energy applications. Part II considered the science and technology of using bioenergy to generate electricity through thermal power plants, and its storage and use via fuel cells and batteries.

Part III takes a closer look at the production and utilisation of the most important liquid and gaseous biofuels in use today. After providing an overview of the nature, properties and use of biofuels on the global scale, it undertakes a detailed study of bioethanol and biobutanol, followed by a similar coverage of biodiesel and biogas. These four fuels are the most significant liquid and gaseous biofuels that are either currently in use or hold promise for the future. It ends by considering the development and use of newer fuels that are now entering the fuel market.

doi:10.1088/978-0-7503-3555-3ch8

This chapter introduces the use of liquid biofuels for transportation by starting with a discussion of the nature and properties of biofuels, especially with respect to their role in emissions reduction and climate change mitigation.

After briefly appraising the global production of biofuels in section 8.3, it turns to a detailed discussion of the properties of transportation engine fuels and the standards and legislations that are used to regulate them. Most internal combustion engines that are in use today were designed to use petrol (gasoline) and petroleum diesel as their energy sources. A thorough knowledge of the fuel standards (e.g. octane and cetane numbers) and legislations is thus needed before biofuels can be used as alternative fuels in such vehicles. These latter fuels differ from petroleum fuels in certain important ways (e.g. oxygen content) and thus require a separate set of standards of their own. These are briefly discussed towards the end of the section. The chapter ends with an Australian case study of transportation fuel standards and legislations.

8.2 Types of biofuels

Chapter 4 showed how lipids could be chemically converted to biodiesel, and how bioethanol, biobutanol and biogas are produced by the biochemical processes of anaerobic respiration and methane fermentation. It also briefly indicated how electrochemical methods could be used to produce hydrogen, the new biofuel that is entering the fuel market. The next three chapters consider in detail the use of these biofuels in the transportation and power generation sectors.

It is natural to begin by considering what biofuels are, why they are used, and the standards and legislations employed to regulate their use.

8.2.1 What are biofuels?

Biofuels are solid, liquid or gaseous substances derived from plant or animal matter. They are used as alternatives to fossil fuels in an endeavour to reduce the greenhouse gas emissions (mainly carbon dioxide) that fossil fuels produce, and which contribute to global warming and climate change.

8.2.1.1 What does carbon neutral mean?

With the exception of hydrogen, all known biofuels release carbon dioxide when they are burnt to release energy. However, the amount of carbon released during the fuel combustion/use is the same as that removed by the biofuel from the atmosphere during its formation through the process of photosynthesis. In this sense, all biofuels are considered **carbon neutral**.

8.2.1.2 Are biofuels truly carbon neutral?

A biofuel will, however, cause the further release of carbon dioxide during the processes of its production (starting at the early stage of land preparation for the biofuel feedstock plantation, to its harvesting, extraction and chemical conversion) and its transportation to its final destination. It is possible that the total amount of such emissions per unit energy produced by the biofuel may exceed that due to the fossil fuel the biofuel replaces. In such a case, the emissions saving becomes negative and the purpose of using the biofuel is defeated. The production and use of biofuels may also be deemed economically unviable if the costs involved in their production and use exceeds those for fossil fuels.

As the net emissions attributed to a biofuel are due to the equipment, materials and processes used in its production, transportation and storage, one must ensure that the contribution due to these factors, per unit energy produced by the biofuel, is kept as low as possible. One way of doing this is to choose a feedstock that has a high energy density (for instance, has a high concentration of ligno-cellulosic compounds).

8.2.2 What are the generations of biofuels?

Biofuels (and the feedstock from which they are derived) are divided into **four generations**, ranked largely according to their ability to reduce carbon emissions as compared to the emissions produced by fossil fuels [1, 2].

First generation biofuels are produced from feedstocks containing high concentrations of sugars, starch or lipids. They typically consist of food crops such as those producing sugar, starch and vegetable oils [1, 2], and provide feedstock for both ethanol and biodiesel production.

Second generation biofuels are produced from non-food biomass with either high ligno-cellulosic content (in the case of ethanol feedstocks) or inedible oil content (for biodiesel production). They include woody biomass such as energy crops, crop residues including wheat straw, corn stover, and inedible oil producing crops such as jatropha, or pongamia. They generally have a higher energy content, which enhances their ability to reduce carbon emissions.

Third generation biofuels are produced from algae (which are often called algae fuel or oilgae for this reason). These are better than second generation biofuels due to their higher energy content [2].

Fourth generation biofuels [3] are derived from genetically engineered plants/biomass that have:

- higher energy yields, or
- low barriers to ligno-cellulosic hydrolysis, or
- can be grown on non-agricultural land.

These feedstocks are produced through genetic engineering of microalgae to enhance biofuel production, leading to much-reduced net emissions (that may even become negative in some cases [4]).

(See exercise 8.1—Generations of biofuels.)

8.2.3 Issues with first generation biofuels

As many of the first generation biofuel feedstocks consist of food crops, the use of such resources for energy production takes place at the expense of food production. This evokes both moral issues as well as logistical ones. The use of food crops for energy production deprives people (including those in poor countries) of their means of food production. This occurs as a result of land used for food production being allocated for energy use. A pertinent example is provided by the production of ethanol from corn plantations in the USA. The dilemma is exacerbated by the tax incentives provided by the government for such re-allocation of land use.

Another issue with first generation biofuels is the relatively low mass fraction of sugars, starch and oils as compared to the total mass of the plant material from which the feedstock is derived. This yields low calorific values as compared to the second generation feedstocks which makes biofuels derived from first generation feedstocks less effective as agents for mitigating greenhouse gas emissions.

The debate on the suitability of ethanol derived from first generation feedstock has been ongoing for a long time. The discussions have centred around the following two questions: Does ethanol make a viable first generation biofuel as a climate change mitigating agent? Are the net emissions from ethanol derived from all first generation feedstocks lower than those from fossil fuels?

It was traditionally believed that bioethanol (E100) fuel reduces carbon emissions, regardless of which feedstock it originated from. The US Department of Energy has maintained the position that corn ethanol was 40% less emitting than gasoline [5].

This is supported by Wang *et al* [6], who found that emissions from ethanol prepared from corn, sugarcane, corn stover, switchgrass and miscanthus could reduce life-cycle emissions by 19–48%, 40–62%, 90–103%, 77–97% and 101–115%, respectively as compared to emissions from gasoline.

But a recent Reuters report on corn-based ethanol by Leah Douglas [7] maintains that the life-cycle emissions of corn-based ethanol is actually higher than that of gasoline. This debate is still ongoing.

8.2.4 Biofuel blends

Biofuels are usually used as fuel blends, and the use of pure biofuel (i.e. E100, B100) is relatively rare. The most common ethanol blends are E10 (10% ethanol, 90% gasoline), E15, E85 (also known as flex fuel) [8]. These blends are used widely in the USA, Brazil as well as other countries.

The most common biodiesel (i.e. Fatty Acid Methyl Ester, or FAME) blends come as B5, B20 and B100 [9]. Note that, with the exception of the E10 blend, the composition of the fuel blends may not be clearly defined, and may vary from country to country. Details for the USA blends are available at [8, 9].

(See question 8.1—Issues with biofuels.)

8.3 Global production of biofuels

Table 1.6 of chapter 1 shows that in 2019, bioenergy (made up of traditional biomass and modern bioenergy) comprised 11.7% of the total global final energy consumption (TFEC). Of this, 6.5% was contributed by traditional biomass, while 5.2% was provided by modern bioenergy (including liquid biofuels).

According to REN21 GSR21 [10] (P93), the three main liquid biofuels used in transportation today are **ethanol**, **biodiesel** (i.e. Fatty Acid Methyl Ester (FAME)) and **hydro-treated vegetable oil (HVO)**. Figures provided by the International Energy Agency (IEA) [11] show that global production of ethanol in 2019 was 115 billion litres. The corresponding figures for biodiesel and HVO were 41 billion litres and 7 billion litres, respectively. The production of ethanol was negatively impacted in 2020 by the onset of the COVID-19 pandemic, which had a severe impact on the global transportation industry, putting serious constraints on the global demand for this biofuel. The production of FAME biodiesel was less affected.

Despite the rapid development of electrics cars (EVs and PHEVs) over the last few years, it is probable that biofuels will continue to play a major role in the production of lower emission transportation fuels for at least the near future. Indeed, the IEA forecasts that global demand for biofuels will grow by 41 billion litres over the 2021–26 period [11].

8.4 Engine fuel properties and standards

The primary purpose of an engine fuel is to provide power to the engine. It must do this in such a way that the engine performs according to its expectations. The fuel must also be stable (i.e. its quality should not degenerate with time). Some of the mandatory criteria that a fuel should satisfy therefore are:

(i) the fuel must provide energy to the engine;
(ii) the engine must perform to its expectations under the use of the fuel;
(iii) the engine must not suffer damage due to fuel use;
(iv) the fuel quality must not deteriorate with time or during storage.

It must be clear from the above considerations that engine fuels need to possess certain properties to qualify as acceptable fuels.

8.4.1 Fuel properties

In the first analysis, the properties that determine fuel quality must include the following:

- **Net Calorific value** (also called its **Lower Heating Value** or Heating Value)—this indicates the energy contained in the fuel per unit mass;
- **Viscosity**—which measures how easily the fuel flows and affects its atomisation properties;
- **Lubricity**—which is essentially a measure of the lubricating action of the fuel on engine parts;
- **Iodine number**—which measures the fuel's stability against chemical decomposition;
- **Flash point**—which is the temperature at which the fuel ignites in the presence of an ignition source;
- **Octane number**—which measures the fuel's resistance to 'knocking';
- **Cetane number**—which is a measure of the time delay between the fuel injection and auto-ignition.

Generally speaking, the net calorific value, lubricity, flash point and iodine number of a fuel should be high, and its viscosity low. The octane and cetane

Table 8.1. Typical values of the fuel properties for (fossil fuel derived) petroleum diesel compared to that of FAME biodiesel used in diesel engines. (CME is Coconut Methyl Ester, obtained from the transesterification of coconut oil with methanol. cSt = centi-Stoke.)

Fuel property	Petroleum diesel	CME	Comment
Cetane number	51	70	Should exceed 50
Flash point	49 °C	106 °C	Should be high to avoid accidental ignition during storage and handling
Viscosity	3–4 cSt	2–3 cSt	Should be low to allow good fuel atomisation in the engine
Lubricity	3800	>7000	Should be high to ensure good lubrication quality
Solvency	Mild	High	Organic particulates should be soluble in the fuel
Total glycerine content	Nil	1.145	Should not exceed 0.24 units

numbers of a fuel will depend on the type of engine, and will thus differ with the type of fuel (see discussion later). Table 8.1 indicates the typical values of some of the above fuel properties for diesel engine fuels.

It must also be noted that the use of fuels invariably results in the production of **undesirable emissions** into the environment. These are harmful to both humans and to the biosphere as a whole. Thus, another criterion that a good fuel must satisfy is that any emissions it produces must be at an acceptable level.

As indicated above, different types of engines require different types of fuels to operate effectively. The fuel requirements for the **spark-ignition engine** (also known as the **gasoline engine**) are thus generally different from those of the **compression-ignition engine** (or **diesel engine**). The spark-ignition engine uses the **Otto cycle**, where the fuel–air mixture needs to be ignited using a spark plug. In the compression-ignition engine, the mixture auto-ignites when the pressure is high enough. To ensure that they operate effectively, these two engines need fuels with physical and chemical properties that are generally different for the two applications. Two of the most important of these properties are the **octane number** and the **cetane number**. Box 8.1 describes these properties in more detail.

(See exercise 8.2—Engine type and fuel properties.)

Box 8.1. Octane number, cetane number and engine type.

Octane number

The octane number of a fuel is a measure of its ability to resist 'knocking' during combustion in a spark-ignition engine. Knocking occurs when the pressure in front of the flame in the cylinder becomes high enough for a secondary (unwanted) ignition to take place. As this ignition is out of synchronisation with the engine cycle, it leads to inefficient fuel use as well as unnecessary wear and tear of the engine.

The quoted octane number (also called octane rating) of a fuel is the mean of the results of two octane rating measurement methods. These are called the **Research Octane Number (RON)** and the **Motor Octane Number (MON).** These are octane numbers determined by using the same method but under different conditions of temperature and pressure.

The octane number is related to the pressure at which the fuel auto-ignites in a test engine [12]. In the testing procedure, the pressure at which a **Primary Reference Fuel (PRF)** auto-ignites in the testing machine is noted. An octane rating of 100 is assigned to the PRF with the best fuel quality (i.e. highest resistance against knocking), and 0 to a chosen PRF with the lowest resistance. The PRF used to determine the high end of the octane number scale is the isomer of **octane (C_8H_{18})** known as **2,2,4 trimethyl pentane**. The PRF used to establish the low-end of the scale is the **normal isomer of heptane, C_7H_{16}**.

In the USA, octane numbers for gasoline engine fuels are [12]:

- 87 for regular gasoline;
- 88–90 for mid-grade gasoline; and
- 91–94 for premium grade gasoline.

Clearly, good gasoline engine fuels are those which have the highest octane numbers.

Cetane number

The cetane number (CN) of a fuel quantifies the delay between the injection of a fuel and its auto-ignition under compression. It measures the fuel quality of a diesel engine fuel, where the fuel–air mixture is ignited through the temperature rise generated through the compression of the mixture. The shorter the delay, the more suitable the fuel is for a diesel engine [13]. Fuels with short delay times are assigned higher cetane numbers. Just as for the case of the octane number, the cetane number of a fuel is determined by experimentally determining the desired property of a fuel (i.e. the delay time), by comparing it with that of standard fuels.The liquid hydrocarbon n-hexadecane ($C_{16}H_{34}$) is an excellent diesel engine fuel and is assigned the cetane number of 100. At the other end of the scale, the hydrocarbon 1-methyl naphthalene ($C_{11}H_{10}$) is a very poor diesel engine fuel and is assigned a cetane number of zero [14]. It must be noted that all engine fuels can be assigned octane and cetane numbers. What is important to note is that the octane numbers are more relevant to quantifying the performance (ignition quality) of a spark-ignition engine, whereas cetane numbers are relevant to diesel engine fuels.

Summary

In summary:

- The octane number of a fuel is used to describe the ignition quality of a spark-ignition (i.e. gasoline) engine fuel. The higher the octane number, the better the fuel.
- The cetane number of a fuel is used to describe the ignition quality of a compression-ignition (i.e. diesel) engine fuel. Fuels with short time delays are good fuels. The shorter the time delay, the higher the cetane number and the better the fuel.
- Good gasoline engine fuels have high octane numbers. Good diesel engine fuels have high cetane numbers.

8.4.2 Gasoline and diesel engine fuels

The main liquid transportation fuels in use today are **gasoline (petrol)** and **petroleum diesel** in the case of land transportation, **jet fuel** for the aviation industry and **diesel/HFO** for power generation and sea transportation. In addition to these fossil-derived fuels, the liquid biofuels **ethanol** and **biodiesel** are used to produce fuel blends such at E10, E85, B5 and B20. **Hydro-treated vegetable oil (HVO)** is rapidly becoming prominent as the preferred alternative to the FAME biodiesel.

Gasoline consists mainly of saturated hydrocarbons called **alkanes** (also called paraffins), which are linear molecules with single bonds between the carbon atoms, the unsaturated hydrocarbons called **alkenes** (also called olefins), which have one or more double bonds between the carbon atoms, and **aromatic hydrocarbons** (where the carbon atoms form ring structures with alternating single and double bonds between the carbons).

Diesel fuel consists mainly of alkanes, aromatic hydrocarbons as well as **napthenes** (where the carbon atoms form ring structures with single bonds only between the carbon atoms [15]). In addition to these molecular mixtures, each fuel may contain a range of impurities and additives that may affect their fuel performance or contribute to environmental emissions. In the case of gasoline, for instance, these may include methanol, ethanol, other oxygenates, benzene, lead or sulphur.

(See question 8.2—Relation between molecular structure and properties of fuels for gasoline and diesel engines.)

8.4.3 Fuel standards and legislations

Any fuel must satisfy both the fuel performance requirements and the environmental performance requirements laid down by the country's transportation and environmental protection legislations. Such standards and legislations are necessary to bring about uniformity in the quality of a country's fuel supply as well as meeting its environmental protection regulations. These legislations are often implemented in the form of **Directives**.

In addition to such regulations, auto manufacturers have specified their own standards recommendations through a World-Wide Fuel Charter (WWFC).

The general performance of the gasoline engine fuel is defined in terms such as its **octane number**, **volatility**, **non-alkane content** and the amount of any **fuel additives**. Its environmental performance is defined in terms such as its **aromatic**, **benzene** and **sulphur** content. In addition, the fuel quality must be tested in order to monitor compliance with the associated standards/legislations. This requires the development of **testing methods**.

A similar set of requirements apply to compression-ignition engine fuels, though the range of values of the properties appropriate for this engine may differ considerably in some cases from those of the gasoline engine.

The world's two major fuel standards-setting organisations/entities are:

- The American Society for Testing Materials (ASTM), which consists of international committees of industry professionals that meet to develop and update standards and test methods for materials, including fuels;
- The European Committee for Standardization (CEN), which develops fuel standards for the EU.

Some examples of gasoline and diesel fuel legislations and standards are:

- The ASTM D4814 standards for gasoline, used in the USA;
- The Fuel Quality Directive 2009/30/EC for gasoline used by the EU;
- The ASTM D 975 specifications for diesel fuel used in the USA;
- The Fuel and Quality Directive 2009/30/EC for diesel fuel used by the EU.

Table 8.2 provides examples of selected fuel property requirements (i.e. standard specifications) for gasoline as stipulated by the EU Directive, together with the corresponding WWFC recommendations.

Table 8.2. Examples of selected fuel standard specifications of gasoline fuel as stipulated by the EU Directive, and the corresponding WWFC Recommendations (data source: [15]).

Fuel property	EU 2009/30/EC specification	WWFC 2013 recommendation
RON	⩾95	⩾95
MON	⩾85	⩾85
Final B.P. (°C)	—	⩽205
Oxygen content (vol%)	⩽3.7	⩽2.7
Methanol (vol%)	⩽3.0	Not permitted
Ethanol (vol%)	⩽10.0	⩽10.0
Isopropyl alcohol (vol%)	⩽12.0	—
Olefins (vol%)	⩽18.0	⩽10.0
Aromatic (vol%)	⩽35.0	⩽35.0
Benzene (vol%)	⩽1.0	⩽1.0
Sulphur content (mg kg^{-1})	⩽10.0	⩽10.0
Lead content (mg l^{-1})	⩽6.0	Not detectable

Table 8.3 provides examples of selected fuel standard specifications for petroleum diesel fuel used in the EU, together with the WWFC recommendations.

Table 8.3. Selected diesel fuel standard specifications used in the EU and the associated WWFC recommendations (data source: [15]).

Fuel property	EU 2009/30/EC specification	WWFC 2013 recommendation
Cetane number	⩾51.0	⩾55.0
Density at 15 °C (kg m^{-3})	⩽845.0	820–840
Viscosity at 40 °C (cSt)	—	2.0–4.0
Flash point	—	⩾55.0
Final B. P. (°C)	—	⩽350
Total aromatics (wt%)	—	⩽15
Sulphur content (mg kg^{-1})	⩽10.0	⩽10.0
FAME content (vol%)	⩽7.0	⩽5.0
Other biofuels	—	HVO, BTL
Methanol//ethanol (wt%)	—	Not detectable
Water content (mg kg^{-1})	—	⩽200
Lubricity at 60 °C (wsd (μm))	—	⩽400

(See exercise 8.3—Separating the engine performance standards from those for environmental performance.)

8.4.4 Biofuel standards

Biofuels have chemical and physical properties that are, in some instances, quite distinct from those of petroleum fuels. An obvious example is that, unlike fossil fuels which are pure hydrocarbons, biofuels contain oxygen as an integral part of their molecular structures (for example, alcohols contain the OH group, and biodiesels are esters). Thus, a new set of fuel standards needs to be stipulated for these fuels.

Table 8.4 gives a list of some of the most important biodiesel quality standards used in the world.

Table 8.4. List of the most important biodiesel standards used by various countries (after Barabas and Todorut [16]).

Country/region	Name of fuel specification	Type of specification
EU	EN14214	Standard for FAME biodiesel
USA	ASTM D6751	Standard for B100
India	ISI5607	Standard for B100
Australia	Fuel Quality Standards (Biodiesel) Determination 2019	Standard for B100

8.4.5 Fuel standards and legislations—the Australian case study

To see how fuel standards and legislations are regulated in a country, it is illustrative to consider a country case study.

The Australian Department of Industry, Science and Energy Resources (DISER) regulates the quality of transportation fuels in the country through the **Fuel Quality Standards Act 2000** and the **Fuel Quality Standards Regulations 2019** [17]. Together with the National Measurements institute, this department:

- sets fuel quality standards and labelling requirements;
- assesses requests to temporarily vary fuel standards; and
- monitors fuel standards compliance.

DISER provides the fuel standards/specifications for petrol, diesel, biodiesel, autogas, ethanol E10, E85 in the form of **Determinations**.

As an example, the petrol fuel standards are described by the **Fuel Quality Standards (Petrol) Determination (2019)** [18]. The standard is provided in the form of a matrix with three columns, showing how each of the 18 fuel parameters (i.e. properties) is specified, and the testing method used to determine the value of the parameter. Table 8.5 reproduces three rows of the standards for petrol.

(See activity 8.1—Engine fuel standards for your own country.)

Table 8.5. A sample of the Australian fuel quality standards for petrol (data source: [18]).

Item	Parameter	Specifications	Testing method
1.	Aromatics	45% v/v maximum with 35% v/v maximum pool average across all grades	ASTM D1319
2.	Benzene	1.0% v/v maximum	ASTM D5580
6.	Ethanol	10% v/v maximum	ASTM D4815

Exercises, questions and activities

Exercise 8.1. Generations of biofuels.

a) What is the primary reason for categorising biofuels into generations?
b) What are two important differences between first and second generation biofuels?
c) In what ways are fourth generation biofuels distinct from earlier generations?

Exercise 8. 2. Engine type and fuel properties.
The paragraphs below relate to the required fuel performance of spark-ignition and compression-ignition engines, and seek to illustrate why the two types of engines require fuels with distinctly different properties. The two paragraphs contain pairs of options separated by hyphens. Circle the correct option that applies to the paragraph.

In a spark-ignition engine, the fuel auto-ignites/is ignited using a spark. The tendency to auto-ignite is measured by the octane number/cetane number. The fuel for a spark-ignition engine must have a high/low tendency to self-ignite. Its octane number should be high/low.

In a compression-ignition engine, the fuel auto-ignites/is ignited using a spark. The time delay between the injection and auto-ignition of the fuel is called the octane number/cetane number. A good compression-ignition fuel should auto-ignite easily/have a high resistance to auto-ignition. The time delay between the injection and ignition of the fuel must be low/high. The cetane number of the fuel must be low/high.

Exercise 8.3. Separating the engine performance standards from those for environmental performance.

Tables 8.2 and 8.3 list both the engine performance and the environmental performance properties for gasoline and diesel engine fuels, respectively. Use a suitable table to group these properties according to these two separate categories, and add a third column to explain why they belong to the indicated category. Do this for each table separately.

You may find that some of the properties belong to both categories.

Question 8.1. Issues with biofuels.

The E10 blend of ethanol has become widely promoted as an environmentally friendly blend for gasoline engine fuels (for reducing both toxic carbon monoxide as well as greenhouse gas carbon dioxide emissions) world-wide. However, recent reports indicate that the GHG emissions-reducing ability of corn ethanol (produced in large amounts in the USA and other countries) is debatable.

The Intergovernmental Panel on Climate Change (IPCC) has made statements about this issue in the early history of the development and use of E10/ unleaded gasoline engine fuels.

(a) Prepare a list of such statements that have been made by the IPCC and/or other scientific institutions (such as the US Department of Energy), and comment on your findings.

(b) The use of second generation ethanol, produced from the hydrolysis of ligno-cellulosic feedstock, would (arguably) alleviate this problem. What is the current state of development of this technology?

Question 8.2. Relation between the molecular structure and properties of fuels for gasoline and diesel engines.

The performance of an engine fuel depends on its physical properties, which in turn are determined by its chemical properties. Both of these are dependent on the molecular structure of the fuel molecules, in particular, the manner in which the atoms are bonded together.

List the engine performance requirements for both the gasoline and diesel engines, and show how they are related to their octane and cetane numbers. Next identify the types of molecules possessed by the known fuels for each type of engine and show how they are related to their octane and cetane ratings.

Activity 8.1. Engine fuel standards for your own country.

What is the engine fuel standards of your own country (if it is other than Australia)? Use the information in box 8.1 above to describe these standards as fully as possible.

If your country is Australia, you may select a different country in the neighbourhood of Australia to carry out this exercise.

References

[1] Demirbas A 2011 Competitive liquid biofuels from biomass *Appl. Energy* **88** 17–28

[2] Sampa S Generation Biofuels. 14.3 Biomass and Types of Biofuels. Science Direct. 2019 [accessed 9 May 2022]. Available from https://sciencedirect.com/topics/engineering/generation-biofuels

[3] Fourth Generation Wix.com [accessed 13 June 2022]. Available from https://allaboutbiofuels.wixsite.com/biofuels/fourth-generation#:~:text=Fourth%20generation%20biofuels%20are%20derived%20from%20specially%20engineered,non-agricultural%20land%20or%20bodies%20of%20water.%20%28Citation%20T-27%29

[4] Radakovits R *et al* 2010 Genetic engineering of algae for enhanced biofuel production *Eukaryot. Cell* **9** 486–501

[5] Alternative Fuels Data Center US DoE. Ethanol Benefits and Considerations [accessed 5 June 2022]. Available from https://afdc.energy.gov/fuels/ethanol_benefits.html

[6] Wang M, Han J, Dunn J B, Cai H and Elgowainy A 2012 Well-to-wheels energy use and greenhouse gas emissions of ethanol from corn, sugarcane and cellulosic biomass for US use *Environ. Res. Lett.* **7** 045905
[7] Environment Reuters. US corn-based ethanol worse for the climate than gasoline, study finds [accessed 13 June 2022]. Available from https://reuters.com/business/environment/us-corn-based-ethanol-worse-climate-than-gasoline-study-finds-2022-02-14/
[8] Alternative Fuels Data Center US DoE. Ethanol Blends [accessed 13 June 2022]. Available from https://afdc.energy.gov/fuels/ethanol_blends.html
[9] Alternative Fuels Data Center US DoE. Biodiesel Blends [accessed 13 June 2022]. Available from https://afdc.energy.gov/fuels/biodiesel_blends.html
[10] REN21.Renewables 2021 Global Status Report. REN21;2021 [accessed12 June 2022]. Available from https://ren21.net/gsr-2021/
[11] Renewables 2021 IEA. Report extract. Biofuels [accessed 12 June 2022]. Available from https://iea.org/reports/renewables-2021/biofuels?mode=transport&region=World&publication=2021&flow=Consumption&product=Ethanol
[12] Gasoline Explained US Energy Information Administration (EIA) [accessed 29 May 2022]. Available from https://eia.gov/energyexplained/gasoline/octane-in-depth.php
[13] Cetane number Energy Education. 2018 [accessed 14 June 2022]. Available from https://energyeducation.ca/encyclopedia/Cetane_number#cite_note-RE2-
[14] DieselNet Technology Guide Fuel Property Testing: Ignition Quality. 2022 [accessed 30 May 2022]. Available from https://dieselnet.com/tech/fuel_diesel_ignition.php
[15] Technology collaboration on Advance Motor Fuels IEA-AMF. Diesel and Gasoline [accessed 14 June 2022]. Available from https://iea-amf.org/content/fuel_information/diesel_gasoline
[16] Barabas I and Todorut I-A 2011 Biodiesel quality, standards and properties. Montero G ed. *Biodiesel – Quality, Emissions and By-Products* (Rijeka, Croatia: Intech Open). Available from https://cdn.intechopen.com/pdfs/23666/InTech-Biodiesel_quality_standards_and_properties.pdf
[17] Australian Government Department of Industry, Science and Energy Resources. Regulating Australian Fuel Quality. May 2022 [accessed 3 June 2022]. Available from: https://industry.gov.au/regulations-and-standards/regulating-australian-fuel-quality
[18] Australian Government Federal Register of Legislation. Fuel Quality Standards Petrol Determination 2019. 2019 [accessed 3 June 2022]. Available from https://legislation.gov.au/Details/F2021C00147/Html/Text#_Toc63780305

IOP Publishing

Anirudh Singh

Chapter 9

Bioethanol and biobutanol

Ethanol and butanol are two biofuels that are suitable as alternative fuels for transportation. They can be produced through the biochemical processes of fermentation of carbohydrate substrates. This chapter considers the fuel properties of these biofuels, and describes their industrial production. The metabolic pathways involved in their production are described in detail. Issues with each fuel are considered, and progress made towards addressing these issues using the tool of metabolic engineering is reported.

9.1 Introduction

Chapter 8 provided a first introduction to biofuels by introducing their nature, the forms in which they are used in transportation and the fuel standards by which they are assessed and regulated globally and regionally.

This chapter takes a detailed look at two biofuels that are produced from the biochemical treatment of biomass—bioethanol and biobutanol. Section 9.2, which follows, begins by considering the fuel properties of ethanol as compared to gasoline, and gives a brief description of its industrial production. This is followed by a detailed review of the metabolic pathway for bioethanol production that leads from the glycolysis of glucose to ethanol fermentation. Recent work, which reveals the presence of two pathways that link the glycolysis to the fermentation, is described. Several issues relating to the use of ethanol as a transportation fuel are enumerated, and progress in research to improve ethanol yield through metabolic engineering is reported.

Section 9.3 considers biobutanol in a similar way. The fuel properties of this biofuel are found to be generally better than those of bioethanol. However, its industrial production is still small compared to that of ethanol. The ABE process for the production of biobutanol is described. This suffers from low butanol yields. The

doi:10.1088/978-0-7503-3555-3ch9

section ends by reporting the results of recent strategies to improve the yields of biobutanol through metabolic engineering.

We start this chapter by looking at bioethanol and its properties as transportation fuel.

9.2 Bioethanol

9.2.1 Fuel properties of ethanol

When considering ethanol as a candidate for transportation engine fuel, it is not sufficient to consider its energy content (i.e. its net calorific value or LHV) only. Other properties relevant to its use in spark ignition engines also become important. It is obviously desirable that these properties are comparable to, if not better than, the corresponding values for gasoline, which is the fuel for which such engines were designed.

A notable feature of gasoline is that it consists of pure hydrocarbons only. In contrast, ethanol is an oxygenated fuel, and it is thus of special interest to see whether this difference in fuel chemistry has any impact on the desired properties. Table 9.1 compares some of the fuel properties of ethanol with that of gasoline.

As seen in the table, ethanol has a lower energy content than that of gasoline and a slightly higher density. This means its volumetric consumption as a gasoline engine fuel is slightly higher as compared to that of gasoline. This disadvantage is offset by the large value of its octane number. The values of both RON and MON are seen to be higher than that of gasoline. The addition of ethanol to gasoline to form a fuel blend assists in increasing the effective octane number and thus increasing the fuel quality of the resultant blend. Other properties lie within acceptable ranges, or are superior to those specified for gasoline (see, e.g. chapter 8, table 8.2).

Table 9.1. Comparison of fuel properties of gasoline (petrol) and ethanol (data source: [1–3]).

Property	Gasoline		Ethanol	
	Data source [1]	Data source [2]	Data source [1]	Data source [3]
Chemical composition	C_4–C_{12} chains	C_4–C_{12} chains	C_2H_5OH	C_2H_5OH
Calorific value (LHV) (MJ kg^{-1})	42.8	42.6	26.8	27.0
Density at 15 °C (kg m^{-3})	742	720–775	790	794
Boiling point (°C at 1 atm)	25–215	≤210	78	78
Auto-ignition temperature (°C)	~300	300	420	423
RON	90	≥ 95	106	108–109
MON	81–89	≥ 85	89	
Cetane number		8–14		2–12
Viscosity at 20 °C (mm^2 s^{-1})		0.83		1.525

9.2.2 Industrial production of bioethanol

Most of the world's ethanol is still produced from first generation feedstocks. The USA produces bioethanol from corn, while Brazil uses molasses, a by-product of sugar production from sugarcane.

The commercial processes used in ethanol production are **dry milling** and **wet milling** [4, 5]. In dry milling, the feedstock (corn kernel in the case of corn) is ground into flour, mixed with water and fermented to produce alcohol. In the wet milling process, the starch, protein and fibre are separated from the feedstock first, and the first of these products used for ethanol fermentation. In the USA, 90% of ethanol is produced using the dry milling process [4].

The dry milling process consists of the following six steps [5]:

(i) **Milling**—the corn kernel is ground into flour.
(ii) **Liquefaction**—water and enzymes are added to form a 'mash' or 'slurry'. The enzyme (α-amylase) breaks down the starch polymer into smaller bits (e.g. dextrin).
(iii) **Saccharification**—the enzyme gluco-amylase is added to break down the dextrin into individual glucose monomers.
(iv) **Fermentation**—yeast bacterium (*Saccharomycese cerevisiae*) is added to convert glucose to ethanol, carbon dioxide and a high protein and fibre co-product called **distillers grains.**
(v) **Distillation**—the fermented product (containing 10–15% ethanol) is distilled to produce hydrous ethanol (a 96% ethanol to 4% water azeotropic mixture). This may be further dehydrated to 100% (anhydrous) alcohol using special techniques.
(vi) **Denaturation**—pollutants (e.g. gasoline) are added to make the alcohol undrinkable.

(See research question 1—Commercial production of ethanol from sugarcane and hydrous ethanol for transportation.)

9.2.3 Metabolic pathways for bioethanol

Ethanol is a metabolic by-product of anaerobic respiration. Ethanol fermentation occurs when oxygen is absent from the reaction environment, and yeast or other microbial hosts are present to provide the necessary enzymes for ethanol fermentation (see chapter 4, section 4.3.3).

Like any other biochemical product, bioethanol is produced through its own specific chain of reactions (i.e. metabolic pathway). Each reaction is catalysed by a specific enzyme, which is provided by a microbial host. This may be a fungus or a bacterium.

Recent research shows that the pathway for ethanol production from a carbohydrate-derived substrate is not unique (i.e. there are at least two variations of the chain of reactions that lead from glucose to ethanol). This implies the presence of alternative enzymes that eventually lead to the final product. The nature of these enzymes needs to be elaborated on, and the identity of the microbial hosts that produce these enzymes needs to be ascertained.

The most common microbial host for ethanol production is yeast (*Saccharomyces cervisiae*), which possesses the enzymes that can catalyse one possible metabolic pathway for ethanol production. The full metabolic pathway for ethanol production starts from the glycolysis of glucose (which produces pyruvic acid as the end product), followed by an ethanol fermentation pathway. The glycolysis part of the pathway converts glucose to the pyruvate anion. This may be described by the overall chemical equation:

$$\begin{aligned} &C_6H_{12}O_6 + 2ADP + 2P_i + 2NAD^+ \rightarrow \\ &2CH_3COCOO^- + 2ATP + 2NADH + 2H_2O + 2H^+ \end{aligned} \tag{9.1}$$

where

- the 6-carbon glucose molecule is converted to the 3-carbon pyruvate anion CH_3COCOO^-;
- ADP is the adenosine diphosphate molecule;
- P_i is the inorganic phosphate anion that is added to ADP to convert it to adenosine triphosphate (ATP);

and the other terms have their usual meanings (see chapter 4 for a review of glycolysis).

It was mentioned in chapter 4, section 4.3.3, that pyruvate (the final product of glycolysis) is converted to ethanol in the process known as ethanol fermentation, but how this occurs was not discussed. It is now known that this fermentation part of the metabolic chain of reactions takes place in two steps:

1. Yeast helps to convert the pyruvate to acetaldehyde and carbon dioxide via the reaction

$$CH_3COCOO^- + H^+ \rightarrow CH_3CHO + CO_2 \tag{9.2}$$

2. The acetaldehyde is then reduced to ethanol with the help of other enzymes (produced by either yeast or some other microbes) via the reaction

$$CH_3CHO + NADH \rightarrow C_2H_5OH + NAD^+ \tag{9.3}$$

But the actual situation is more complex. Recall that in the case of aerobic respiration, a co-enzyme called **acetyl CoA** played a part at the end of the glycolysis stage. Does it play a part in anaerobic respiration and ethanol fermentation path as well?

It is found that there are two possible ways of converting the pyruvate anion to acetaldehyde [6]:

(i) a two-step process that does not involve acetyl CoA;
(ii) a three-step process that uses acetyl CoA as an intermediary between the glycolysis and fermentation chains of the full metabolic pathway for ethanol fermentation.

The two-step process is what was described above (through equations (9.2) and (9.3)). In the three-step process:

(i) the pyruvate is first converted to acetyl CoA through a process of decarboxylation (i.e. removal of the carboxylic acid functional group);
(ii) the acetyl CoA is converted to aldehyde;
(iii) the aldehyde is reduced to ethanol.

Figure 9.1 shows how these two separate processes contribute to the pathway for ethanol fermentation that starts from the glycolysis of glucose.

With the inclusion of the three-step process in the metabolic pathway for ethanol production, more reactions are added to the chain of reactions forming the pathway. As each reaction must be accompanied by its own enzyme, which is produced by a particular microbe, it is possible that other microbes are involved in the fermentation process. Some of these are known, while others are still the subject of active research. Table 9.2 summarises what is known at the time of writing about the enzymes that facilitate the reactions in the two-step (pathway 1) and three-step (pathway 2) processes mentioned above.

The microbes that produce the enzyme PDC for the two-step process are *S. cerevisiae* (yeast, which is a fungus), *Zymomonas mobilis* and *Sarcina ventriculi*

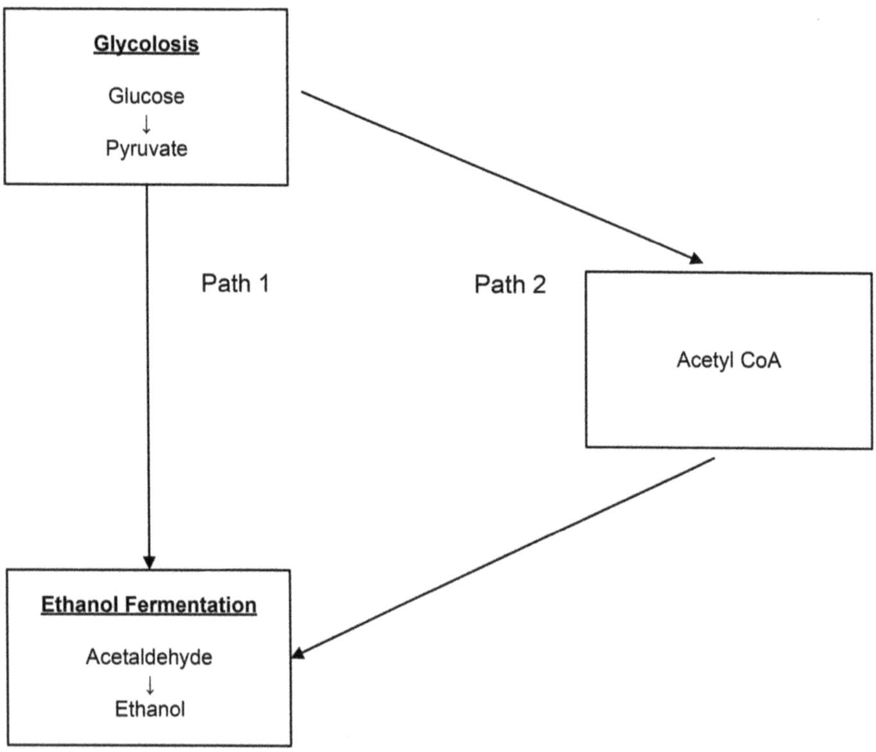

Figure 9.1. A schematic diagram of the alternative metabolic pathways for ethanol production that includes the two possible ways of converting pyruvate into acetaldehyde.

Table 9.2. Known enzymes associated with the two-step (pathway 1) and three-step (pathway 2) processes linking glycolysis to the fermentation of ethanol (data source: [6]).

Steps	Enzymes for pathway 1 (two-step process)	Enzymes for pathway 2 (three-step process)
Pyruvate–acetaldehyde	Pyruvate decarboxylase (PDC)	N/A
Pyruvate–acetyl CoA	N/A	Pyruvate formate lyase (PFL) and pyruvate ferredoxin oxidoreductase (POR)
Acetyl CoA–acetaldehyde	N/A	Acetaldehyde dehydrogenase (AcDH)
Acetaldehyde–ethanol	Alcohol dehydrogenase (ADH)	Alcohol dehydrogenase (ADH)

(which are bacteria), as well as a host of other bacteria and fungi. (The reader is referred to the paper by Eram and Ma [6] for further details.)

The situation is complicated by the realisation that microbes having the appropriate enzymes for ethanol fermentation can exist within different temperature regimes (for instance, those ranging from the mesophillic to thermophilic and beyond). As we are interested in maximising yields in the commercial production of ethanol, it is possible that the most appropriate temperature ranges of interest may lie well beyond the ambient temperature. The subject is still being actively investigated.

9.2.4 Issues with bioethanol fuel

There are several issues with bioethanol as a gasoline engine fuel. Some of these are:

(i) Ethanol has a low heating value as compared to gasoline (see table 9.1).

(ii) It forms an azeotropic mixture with water when the mixture reaches a concentration of 96% ethanol. This means that distillation is no longer effective in increasing the ethanol concentration, and special (more expensive) dehydration techniques have to be used to obtain anhydrous (i.e. water-free) ethanol.

(iii) Ethanol is hygroscopic, and thus absorbs water from the atmosphere spontaneously when left alone.

(iv) Ethanol becomes toxic to yeast at a concentration of ~10%. This means that the fermentation process stops when this concentration is reached, and one needs to resort to distillation to obtain higher levels of concentration.

As seen above, these issues are related to both the production of ethanol as well as to its fuel properties. One of the ways in which the former can be addressed is through the use of genetic engineering.

9.2.5 Future biofuels through metabolic engineering

Genetic engineering provides an effective tool for the improvement of the yield of biofuels obtained through fermentation. The metabolic pathway that produces biofuel is determined by the enzymes produced by the host microbe. It may be possible to produce a new strain of the microbe (through genetic engineering) that has enzymes that are more conducive towards producing the biofuel.

Some of the challenges faced in bioethanol production include:

- Fermentation of bioethanol produces many by-products, which compete with the main product (bioethanol) for the available biomass (i.e. the glucose content of the feedstock). This reduces the share of the mass available for the production of the bioethanol.
- The selected host microbe utilises the available biomass inefficiently, or is unable to metabolise many of the hydrolysates available from the feedstock. This occurs in the case of second generation feedstock, which is ligno-cellulosic in nature.
- The end products are toxic to the host, thus stopping the metabolic process before the sugars are used up.

Considering the particular case of ethanol, one first notes that so far this biofuel has been mainly produced from first generation feedstocks (such as sugarcane and starchy crops), which have high glucose content, using yeast as the microbial host. Strategies for improving the ethanol yield include [7]:

- identifying better enzymes/microbial hosts;
- eliminating pathways that compete with ethanol production (i.e. removing pathways that produce the by-products of ethanol fermentation);
- optimising host (microbial) strains for better glucose uptake;
- using better feedstocks (e.g. second generation (ligno-cellulosic biomass) instead of first generation feedstocks).

Some examples of how this can be achieved in practice are:

Case A: First generation feedstocks:

(1) *S. cerevisiae* (yeast) has been traditionally used to produce ethanol from first generation feedstocks. This host uses the Embden–Meyerhof–Parnas (EMP) pathway for the glycolysis of glucose to produce pyruvate. But an alternative host *Z. mobilis* uses the Entner–Doudoroff (ED) pathway instead, and increases ethanol productivity by a factor of 2.5 [7]. It does this by making more of the carbon available from the sugar for ethanol production.

(2) Yeast can be modified (via genetic engineering) to produce a new strain that contains an intra-cellularly localised sucrose invertase (iSUCI) enzyme that increases the ethanol yield by 11% [8].

Case B: Second generation (ligno-cellulosic) feedstocks:

One of the requirements for the production of alcohols from second generation feedstocks is the availability of microbial hosts that provide the enzymes capable of metabolising the products of hydrolysis (i.e. hydrolysates) of ligno-cellulosic materials. One such host that is now available is a special strain of *Z. mobilis* which has been genetically engineered to co-utilise glucose, mannose and xylose (i.e. hydrolysates of first and second generation feedstocks), thus facilitating ethanol production from second generation feedstock. A review of ethanol production from ligno-cellulosic biomass (LCB) is given by Geddes *et al* [9].

(See exercise 9.1—Strategies for improving bioethanol yield.)

9.3 Biobutanol

Butanol is a 4-carbon alcohol occurring as four straight-chain or branched-chain isomers [10]. These consist of:

- n-butanol (1-butanol), which is the straight-chain isomer with the hydroxyl group at the C1 position;
- sec-butanol (2-butanol), the straight-chain isomer with the hydroxyl group at the C2 position;
- iso-butanol, a branched-chain isomer with the hydroxyl group at the C1 position; and
- tert-butanol, which has a branched-chain structure with a central carbon atom bonded to 3 carbons and the hydroxyl radical in a tetrahedral configuration.

n-butanol and iso-butanol are of interest as fuels for transportation.

(See exercise 9.2—The isomers of butanol.)

Butanol holds two important advantages over ethanol as a gasoline engine fuel in that it has a higher LHV, but unlike ethanol, is not hygroscopic and is thus unlikely to absorb water from the atmosphere.

9.3.1 Fuel properties of butanol

The fuel properties of butanol are compared with those of ethanol and gasoline by Mack *et al* [11], and are reproduced in table 9.3.

The table shows that, like ethanol, both n-butanol and iso-butanol have acceptable values of octane number (the average value of RON and MON), low cetane number, high self-ignition temperatures and solubilities in water, and low viscosities. Unlike ethanol, butanol is seen to have a higher heating value that is

Table 9.3. Comparison of fuel properties of butanol, ethanol and gasoline (data source: [11]).

Fuel property	n-butanol	Iso-butanol	Ethanol	Gasoline
Energy content (LHV) (MJ L^{-1})	26.9	26.6	21.4	30–33
Density (kg m^{-3})	809.5	801.8	789.3	720–780
RON	96	105	130	88–98
MON	78	94	96	80–88
Cetane number	25	—	8	0–10
Boiling point (°C)	117.7	108	78	35–200
Self-ignition temperature (°C)	343	415.6	434	~300
Solubility in water (wt%)	7.7	8.7	miscible	negligible
Kinematic viscosity at 20 °C (CSt)	3.6	8.3	1.5	0.37–0.44

closer to that of gasoline, thus assuring better fuel economy for gasoline (petrol) powered vehicles.

9.3.2 Industrial production of biobutanol

Currently, butanol is produced from the same (first generation) feedstock as bioethanol, including corn, grain, etc. [12]. According to Statistica [13], the global market volume of n-butanol in 2020 was 5.12 million tonnes. This is a small fraction of the estimated 81 million tonnes of bioethanol that was produced in the same year.

9.3.3 Butanol fermentation through the ABE process

Butanol was first produced biochemically using clostridium by Louis Pasteur in 1861 [15]. It can be produced by both chemical and biochemical processes. An example of chemical production is the **oxo synthesis process** [10], while the best known biochemical process is the **acetone–butanol–ethanol (ABE) process** involving clostridium bacteria as the microbial host.

During World War I, the ABE process was employed primarily to produce acetone, needed to manufacture cordite for the war effort. In the 1950s, the petrochemical process was used for the production of n-butanol commercially for the first time, and the ABE process was temporarily set aside. However, rising oil prices quickly brought this pathway back into favour.

The ABE process is commonly divided into two consecutive phases, consisting of the **acidogenesis** and the **solventogenesis phases** [10, 14]. The first phase produces acids, which are among the intermediate products of the process. It lowers the pH of the medium, allowing the switch to the next phase. The solvents (acetone, butanol and ethanol) are produced in this (appropriately named) solventogenesis phase.

The metabolic pathway for the ABE process starts with the familiar glycolytic conversion of glucose to pyruvate via the EMP pathway followed by the conversion of this product to acetyl CoA.

Starting from an assumed first generation feedstock such as starch, the ABE pathway may be divided into the following steps:

- the feedstock is hydrolysed to glucose by amylase enzyme;
- the glucose is converted to pyruvate via the EMP pathway;
- the pyruvate is converted to acetyl CoA by the enzyme pyruvate ferredoxin oxidoreductase (POR) [10];
- this acetyl CoA provides the common precursor for all the intermediate and end products formed in the subsequent acidogenesis and solventogenesis phases.

The **acidogenesis phase** is the period in which there is exponential cell growth, resulting in the production of acidic intermediate products **acetate** and **butyrate** (these names deriving from the anions of the acetic and butyric acids that are actually formed). The acetate results when acetyl CoA is catalysed by the enzymes **phosphotransacetylase (PTA)** and **acetate kinase (AK)**. The formation of butyrate involves the conversion of two molecules of acetyl CoA to the 4-carbon **acetoacetyl CoA**, which acts as the precursor for the formation of acetone, butanol and butyrate.

The drop in the pH of the medium causes a switch to the **solventogenesis phase**, thus named because of the formation of the three **solvents**, acetone, butanol and ethanol, during this phase. It involves the re-assimilation of the acetate and butyrate intermediates formed in the first phase.

The reader is referred to [10] for the details of the complete process and a diagrammatic representation of the entire metabolic pathway. Note carefully that for each biochemical reaction there is a dedicated enzyme (only a few of which have been mentioned above).

(See exercise 9.3—Butanol pathways and enzymes.)

9.3.4 Improving yield through metabolic engineering

The ABE process as originally envisioned had several drawbacks.

A few of these are [14]:

- high cost of substrate;
- low butanol yield;
- low cell density and growth rate;
- high energy consumption in butanol recovery from solution.

Li *et al* [14] review the current advances (as in 2020) in enhancing the ABE process in the clostridium species via:

(i) elucidating the metabolic pathways of butanol synthesis;
(ii) use of metabolic engineering to enhance microbial (cellular) performance;
(iii) optimising the ABE process of butanol fermentation.

They assert that while micro-organisms belonging to the *Clostridium* genus can select from 14 or more enzymes that can hydrolyse (i.e. degrade) carbohydrate polymers, only four species give significant results in the synthesis of butanol. These are:

- *C. acetobutylyicum;*
- *C. beijerinckii;*
- *C. saccharoperbutylacetonicum*; and
- *C. saccharoacetobutyllicum*.

They proceed to review the work done in developing successful genetic engineering tools to achieve the goals listed above.

Exercises and questions

Exercises

Exercise 9.1. Strategies for improving bioethanol yield.

a) Associate each of the two examples given for the improvement of ethanol yield with the strategies given in the bullet points. Present your answer in the form of a matrix.

b) By referring to Kang and Lee [7] or otherwise, identify other methods for increasing ethanol yield and add them to your matrix.

Exercise 9.2. The isomers of butanol.

By examining table 9.1, p 176 of Liu *et al* [10], or otherwise, identify the exact location of the OH group in each of the possible isomers of butanol and draw the corresponding structural formula for the isomer.

Exercise 9.3. Butanol pathways and enzymes.

a) The biochemical pathway for the production of butanol is given by Ndaba *et al* [15] in their figure 1, and by Liu *et al* [10] in figure 3. The latter contains the enzymes responsible for each chemical reaction of the pathway.

Draw a sketch of figure 9.1 from Ndaba *et al* [15] and add the names of the enzymes at each chemical step obtained by referring to figure 3 of Liu *et al* [10].

b) This figure shows starch as the starting point of the ABE metabolic pathway. This can be obtained from a first generation feedstock such as corn.
   (i) Comment on the additional steps that need to be taken if the feedstock was a second generation feedstock such as woody biomass.
   (ii) Identify enzymes and the microbial hosts that produce them that could assist in the hydrolysis of the second generation feedstock.

Questions

Research question 9.1. Commercial production of ethanol from sugarcane and hydrous ethanol for transportation.

Carry out a literature search of the commercial production of ethanol from sugarcane and answer the following questions:

(i) What are the by-products produced from this feedstock for ethanol production?
(ii) Is it possible to use hydrous ethanol as a transportation fuel? Do a survey of the literature on ethanol-producing countries and ascertain the nature and composition of ethanol used by them as transportation fuels.

References

[1] Qi D H and Lee C F 2016 Combustion and emission behaviour for ethanol-gasoline blended fuels in a multipoint electronic fuel injection engine *Int. J. Sustain. Energy* **35** 323–38

[2] Technology collaboration on Advanced Motor Fuels IEA-AMF. Diesel and Gasoline [accessed 14 June 2022]. Available from https://iea-amf.org/content/fuel_information/diesel_gasoline

[3] Technology collaboration on Advanced Motor Fuels IEA-AMF.Ethanol properties. [accessed 3 July 2022]. Available from https://iea-amf.org/content/fuel_information/fuel_info_home/ethanol/e10/ethanol_properties

[4] Alternative Fuels Data Center US Department of Energy. Ethanol production and distribution [accessed 3 July 2022]. Available from https://afdc.energy.gov/fuels/ethanol_-production.html

[5] Lets talk science How ethanol is made [accessed 2 July 2022]. Available from https://letstalkscience.ca/educational-resources/backgrounders/how-ethanol-made

[6] Eram M S and Ma K 2013 Decarboxylation of pyruvate to acetaldehyde for ethanol production by hyperthermophiles *Biomolecules* **3** 578–96

[7] Kang A and Lee T S 2015 Converting sugars to biofuels: ethanol and beyond *Bioengineering* **2** 184–203

[8] Basso T O, de Kok S, Dario M, do Espirito-Santo J C, Müller G, Schlölg P S, Silva C P, Tonso A, Daran J-M M and Gombert A K 2011 Engineering topology and kinetics of sucrose metabolism in *Saccharomyces cerevisiae* for improved ethanol yield *Metab. Eng.* **13** 694–703

[9] Geddes C C, Nieves I U and Ingram L O 2011 Advances in ethanol production *Curr. Opin. Biotechnol.* **22** 312–9

[10] Liu H, Wang G and Zhang J 2012 The promising fuel – biobutanol ed Z Fang *Liquid, Gaseous and Solid Biofuels – Conversion Technologies* (London: IntechOpen) p 554 Available from https://intechopen.com/books/3526

[11] Mack J H, Schuler D, But R H and Dibble R W 2016 Experimental investigation of butanol isomer combustion in HCCI engines *Appl. Energy* **165** 612–26

[12] Alternative Fuels Data Center US Department of Energy. Biobutanol. [accessed 3 July 2022] Available from https://afdc.energy.gov/fuels/emerging_biobutanol.html

[13] Statistica Market volume of n-butanol world-wide from 2015–21 [accessed 3 July 2022]. Available from https://statista.com/statistics/1245211/n-butanol-market-volume-worldwide/

[14] Li S, Huang L, Ke C, Pang Z and Liu L 2020 Pathway dissection, regulation, engineering and application: lessons learnt from biobutanol production by solventogenic clostridia *Biotechnol. Biofuels* **13** 1–13

[15] Ndaba B, Chiyanzu I and Marx S 2015 n-Butanol derived from biochemical and chemical routes: a review *Biotechnol. Rep. (Amst.)* **8** 1–9

IOP Publishing

Anirudh Singh

Chapter 10

Biodiesel, biogas and biomethane

This chapter continues the discussion of the science and technology behind the production and use of liquid and gaseous biofuels by considering the cases of biodiesel and biogas. The nature and properties of FAME biodiesel, the present renewable alternative to diesel engine fuel, is discussed and the industrial-scale production of this biofuel is described in full, from the feedstocks used to the range of transesterification techniques employed in its manufacture. The gaseous biofuels biogas and biomethane (also known as renewable natural gas since it can replace LNG as a transportation fuel) are next considered as fuels for heating, power generation and transportation. Biomethane is mostly produced from biogas, and the cleaning/scrubbing and carbon dioxide removal techniques used to perform this fuel upgrading are described.

10.1 Introduction

Biodiesel, biogas and biomethane are all market-tested fuels currently in routine use in the transportation sector. Biodiesel is used as a (blended) substitute to petroleum diesel. Biogas may be purified to produce biomethane, whose properties are very similar to those of natural gas and can thus be used as a 'drop in' fuel substitute in all its applications. This chapter aims to provide a detailed introduction to the properties, production and use of these three biofuels.

Section 10.2 begins by considering the nature and fuel properties of biodiesel. It next describes the industrial-scale production of FAME biodiesel, starting from the nature of the feedstock and its pre-treatment before transesterification. The range of transesterification methods, including base and acid catalysis, biological catalysis using enzymes, as well as supercritical transesterification are introduced.

Section 10.3 introduces biogas and biomethane (also known as renewable natural gas (RNG)), the two main gaseous biofuels available for heating, cooking, power

doi:10.1088/978-0-7503-3555-3ch10

generation and transportation. Biogas is 45–75% methane, the rest being made up of carbon dioxide, water and other (possibly toxic) impurities. It can be used, with minor cleaning and scrubbing, as a fuel for heating, cooking, power generation and transportation. It can also be upgraded to RNG by removing the carbon dioxide component through upgrading. The techniques for cleaning and upgrading of biogas are described in this section. These include:

- primary treatment to remove moisture and particulate matter from raw biogas;
- secondary treatment to remove harmful impurities such a siloxanes, VOCs and hydrogen sulphide; and
- advanced treatment to remove carbon dioxide and other gases.

The technologies for carbon dioxide removal are enumerated.

The section ends with a discussion of the future supply outlook for biogas and biomethane on the global scale.

10.2 Biodiesel

10.2.1 Introduction

In chapter 4, we saw that nature provides high energy content substances in the form of lipids, which appear in two forms: simple or complex. Simple lipids consist of oils, fats and waxes, all of which belong to the class of organic compounds called triglycerides. These are examples of esters, which are formed when an alcohol reacts with a fatty acid.

Figure 4.1 shows a simple example of an ester. This is methyl acetate, formed when the 2-carbon acetic acid (which is a fatty acid) reacts with methanol. A triglyceride is a substance formed when three fatty acids react with glycerol, which is an alcohol with three hydroxyl groups. As natural fatty acids are very large (i.e. have many carbon atoms in a molecule) a triglyceride is made up of very large molecules, thus making it very viscous. Such molecules can be reduced in size through the process of transesterification, where each triglyceride molecule forms three (smaller) ester molecules. Figure 4.2 shows how this process takes place.

After briefly discussing the fuel properties of biodiesel, the following sub-sections of this chapter discuss the processes involved in the transesterification of naturally-occurring feedstocks on an industrial scale, and draw attention to some of the issues of this fuel.

10.2.2 Biodiesel as a fuel

Biodiesel is one of the alternative fuels for diesel engine vehicles and power plants. It must thus satisfy the fuel standards appropriate for diesel engines. These were discussed in general in chapter 8. The standards that are specific to biodiesel will be discussed below in section 10.2.3.

Biodiesel is usually used as a blend with petroleum diesel. These blends were briefly highlighted in chapter 8. The fuel is rarely used in the neat form for vehicles [1].

It is important to observe that the molecular composition of biodiesel is not unique. Unlike other biofuels such as ethanol and butanol, the nature and distribution of molecules making up biodiesel will depend on the feedstock used. This is primarily because the fatty acids that go towards producing biodiesel come from natural sources (the feedstocks) which contain distinctive mixtures of triglycerides. Each biodiesel is thus characterised by its own specific distribution of fatty acids. Representative distributions for some common biodiesel feedstocks are shown in table 4.1. From this table, it should be obvious that the biodiesel derived from soybean oil will have a distribution of fatty acid chains that are quite different from that derived from, for example, corn oil.

These specific distributions of fatty acid chains in the oils and fats that provide the feedstock for biodiesel become an important consideration in the production of HVO, the other alternative fuel for diesel engines (see the next chapter).

10.2.3 Fuel properties of biodiesel

As pointed out in chapter 8, the difference in chemical composition between biodiesel (which contains esters) and petroleum diesel (which is a mixture of hydrocarbons) requires the assignment of a new fuel standard specification to this biofuel. In the USA this standard is ASTMD6751 and its later derivatives, while the standard used by the EU is EN14214 and its later versions. Table 10.1 compares selected fuel standards chosen from these two fuel specifications for biodiesel, as given by DieselNet [2].

Table 10.1. Comparison of the US standards ASTMD6751-12 for biodiesel with the EU standards EN14214:2012 (data source: [2]).

Fuel property	ASTMD6751-12	EN14214: 2012
Flash point (min)	93 (°C)	101 (°C)
Kinematic viscosity	1.9–6.0 (mm^2 s^{-1})	3.5–5.0 (mm^2 s^{-1})
Density	—	860–900 (kg m^{-3})
Ester content (%) (minimum)	—	96.5
Sulphur (maximum)	15 ppm (for S15); 0.05% (for S500)	10.0 mg kg^{-1}
Cetane number	47	51.0
Iodine value (maximum)	—	120 g Iodine/100g
Linoleic acid methyl ester (maximum)	—	12.0% (w/w)
Polyunsaturated methyl ester (maximum)	—	1.00%
Mono, di and triglyceride (maximum)	Mono: 0.40% (w/w)	Mono: 0.7%; di: 0.2%; tri:0.2%
Free glycerol (maximum)	0.02% (w/w)	0.02% (w/w)
Total glycerol (maximum)	0.24% (w/w)	0.25% (w/w)

10.2.4 Industrial production of biodiesel

The essential steps in the industrial production of FAME biodiesel are:

- selection of feedstock;
- oil extraction;
- pre-treatment of feedstock;
- transesterification.

10.2.4.1 Feedstocks for biodiesel production

Some 350 possible biodiesel feedstocks have been identified [3]. Of these, the following plant-derived feedstocks have been identified as commercially viable:

Palm, jatropha, rapeseed, soybean, sunflower, cottonseed, safflower and peanut oil.

The feedstocks may be broadly divided into [3, 4]:

- edible oils,
- non-edible feedstocks,

where the non-edible feedstocks consist of

- non-edible plant-derived oils,
- algae,
- used cooking oil (UCO), yellow grease, brown grease and soapstock.

Edible oils include palm, coconut, soybean, sunflower, rapeseed and peanut oils. Malaysia and Indonesia use palm and coconut oils. In the USA, soybean is the preferred feedstock, while the EU uses rapeseed. Edible oils are easier to extract from the plant matter, and do not require any chemical processing. However, they are part of the food-vs-fuel debate, and are more expensive to produce than non-edible oils [4].

The non-edible feedstocks include the list of plants consisting of jatropha (*Jatropha curcas*), pongamia (*Pongamia pinnata*), nagchampa or dilo (*Calophyllum inophyllum*), rubber seed tree (*Hevea brasiliensis*) and neem (*Azadirachta indica*). Pongamia and jatropha are especially oil-rich (with seed kernels possessing in excess of 29 (wt)% oil [3]).

Algae are prolific sources of non-edible oil for biodiesel production. They yield more than 100 times more oil per hectare than plant-based feedstock, and are fast-growing.

Used cooking oil, yellow grease and soapstock provide cheap non-edible feedstock. But in common with the plant-derived non-edible oils, these suffer from high free fatty acid concentrations, and require pre-treatment before the transesterification process can begin [4].

10.2.4.2 Methods of oil extraction

The main methods employed for the extraction of oil from feedstock are [3]:

- Mechanical—use of an engine-driven screw press or manual ram press, yielding 60–80% of the total available oil;

- Solvent (i.e. chemical) extraction—this includes soxhlet extraction, ultra-sonification or hot-water extraction;
- Enzymatic—oil seeds ground to reduced particle size are extracted using suitable enzymes.

10.2.4.3 Pre-treatment of feedstock

Pre-treatment refers to all aspects of treatment of the feedstock (including de-hulling, etc.) to prepare it for transesterification. Our main interest is in the chemical pre-treatment of the feedstock.

Many non-edible oils contain free fatty acids (FFAs) and water. The FFAs are present in non-edible oils in abundant amounts. They are formed from the hydrolysis of triglycerides, and will react with the catalyst in base-catalysed esterification to form soap (RCOONA) and water, slowing the reaction down. There is thus a need to remove the water and reduce FFA to less than 0.05 (wt%) before the feedstock is acceptable for transesterification.

Two methods of reducing FFAs in the feedstock [5] are:

(i) Caustic stripping—the use of an alkali to remove the acid via soap formation;
(ii) Acid esterification—addition of an alcohol (e.g. methanol) to remove the FFA through esterification.

(See also Chai *et al* [6].)

10.2.4.4 Methods of transesterification

Transesterification reactions are reversible reactions. This implies that, unless special steps are undertaken, the reaction will be incomplete if a stoichiometric mixture of the reactants is used. One method of increasing yields that is commonly employed is to use an excess of alcohol. Catalysis assists by increasing reaction rates so that greater yields are obtained in shorter time intervals.

Homogeneous and heterogeneous catalysis:

Catalysis for the transesterification of triglycerides can be either **homogeneous** or **heterogeneous**. In homogeneous catalysis, the substrate (i.e. the triglyceride) and the catalyst are in the same phase (usually the liquid phase). In heterogeneous catalysis, the two are in different phases (e.g. the catalyst may be solid while the substrate is in the liquid form).

The advantage of homogeneous catalysis is that high yields are possible at low temperatures due to good miscibility. The major disadvantage is the difficulty in recovering the product from the solution after the completion of the reaction. Here, heterogeneous catalysis (e.g. use of solid catalysts) is able to avoid the separation issue.

The catalysis of the transesterification process may be divided into **chemical** and **biological (enzymatic) catalysis** [7].

Chemical catalysis uses:

- homogeneous catalysts, which may be either alkalis or acids;
- heterogeneous catalysts, comprising solid alkalis, acids, or nanocrystals; as well as
- supercritical fluids, the use of which removes the need for catalysts.

Biological catalysis involves the use of enzymes, which can consist of:

- free lipases;
- traditional immobilised lipases; or
- lipases immobilised on magnetic nanoparticles [7].

Base-catalysed transesterification:

In base-catalysed transesterification, solutions of either NaOH or KOH are used as catalysts. Such homogeneous catalysts produce good yields at low temperatures. Complications arise, however, if the feedstock is a non-edible oil or used cooking oil. As the concentrations of free fatty acids (FFA) in these are typically higher than 0.5%, saponification results from the reaction between the FFA and the alkali, resulting in a decrease in catalyst content and the slowing down of the transesterification reaction [8].

The presence of water as impurity in the reaction medium is also deleterious to the transesterification process in the case of base catalysis. At high temperatures, water is able to hydrolyse triglycerides to produce diglycerides and FFAs. The latter contributes to a reduction in yield due to saponification.

Acid-catalysed transesterification:

In acid-catalysed transesterification, strong acids are used as catalysts instead of alkalis. As alkali metal cations are no longer present in the reaction environment, the use of acid catalysts avoids the issue of saponification. The favoured acid is sulphuric acid, while other inorganic acids such as HCl, BF_3 and H_3PO_4 are also considered [8]. The yields under acid catalysis are higher than those due to base catalysis. However the reaction rates are low, which necessitates longer reaction times at higher temperatures.

Supercritical transesterification:

If the temperature in the biodiesel reactor is above the critical temperature T_c and the pressure exceeds the critical pressure P_c, the boundary between the liquid and gaseous phases of the medium disappears. Therefore, beyond the point (P_c, T_c) (called the **critical point**) in the phase diagram, the distinction between the liquid and gaseous phases also disappears [8].

The region above and to the right of the critical point in the diagram is known as the **supercritical region**. Here, the fluid exhibits both liquid and gaseous properties, producing excellent miscibility amongst the reacting components. Reactions carried out in this (high temperature, high pressure) region of the phase diagram give high yields in short times **without the need for catalysts**. This technique of achieving high yields without the assistance of catalysts was pioneered by Saka and Kushdiana (2001), who produced a 95% yield of canola oil biodiesel at 350 °C and 45–65 MPa pressures in a small reactor within a time of 4 min [9].

Biological catalysis (use of enzymes):

In enzymatic esterification, the chemical (i.e. inorganic) catalyst is replaced by a biological catalyst (i.e. an enzyme). As mentioned above, such enzymes consist of lipases, which may be **free**, **traditionally immobilised**, or **immobilised on magnetic nanoparticles (MNPs)**.

A lipase is an enzyme, found in microbes as well as the organs of animals, that helps in the hydrolysis of triglycerides into glycerol and the component fatty acid molecules. They are present both within cells (i.e. are intra-cellular) or outside cells (i.e. are extra-cellular). Enzymes are a better choice for catalysis because they are biocompatible, biodegradable and environmentally acceptable [7]. The microbial lipases (that can be used in biodiesel production) can be obtained through the use of a large range of lipase-producing microbes such as *Aspergillus niger*, *Candida rugose* and *Fusarium heterosporum*. While considerable research has been carried out on such catalytic agents, they are currently not being used for large-scale biodiesel production. The main challenges encountered in their commercialisation are high costs and the difficulty of enzyme recovery from the solution after product formation.

Lipases can be used as free lipase, which are molecules that move about freely in the medium. Their useful properties such as activity, stability and specificity can, however, be improved by immobilising them on solid materials. These can be immobilised on non-magnetic materials, which give rise to **traditionally immobilised lipases**, or on **MNPs**. In both cases, the yields generally improve (see [7] and references therein for more details and examples).

10.2.5 Issues with biodiesel

The major issues with FAME biodiesel were first highlighted in chapter 4 (see section 4.2.2). FAME biodiesel is an ester, and thus contains oxygen, which is absent in petroleum diesel. Thus, a new set of standards is needed to specify its fuel performance.

As was mentioned in chapter 4, one way of overcoming this issue is to remove the oxygen. Figure 4.3 depicts the two alternative chemical pathways from a vegetable oil feedstock: one to FAME biodiesel and the other to the new alternative fuel for diesel engines known as hydro-treated vegetable oil (HVO). This new fuel will be discussed in chapter 11.

10.3 Biogas and biomethane

As we saw in chapter 6 (section 6.5.4), biogas is well known for its use in the generation of power and heat in CHP plants. When upgraded to biomethane, this energy source can also power transportation vehicles. This section considers the production and use of these two fuels in more detail.

10.3.1 Biogas

10.3.1.1 Feedstock and sources

Raw biogas consists of some 45–75% methane, with the rest being made up principally of carbon dioxide and smaller fractions of impurities [10]. In chapter 4 we saw that this gaseous energy source is produced by groups of bacteria through the process of anaerobic digestion of organic feedstock followed by methane fermentation.

These feedstocks come from a range of organic sources, consisting of both fresh (mostly plant) matter as well as municipal waste products. The World Biogas Association provides detailed information on the following feedstocks: livestock

manure, sewage, food waste (including municipal solid waste (MSW)), crop residues and energy crops (see WBA Report [11]).

The specific feedstocks listed above are used in the controlled production of biogas in biodigesters. In practice, biogas is also produced naturally at certain sites and locations. In the USA, the sources of biogas include [12]:

- landfills (including MSW);
- livestock operations;
- wastewater treatment;
- other sources.

10.3.1.2 Global production

Europe, China and the USA account for 90% of the global production of biogas [10]. Of these, the EU is the largest producer, with Germany contributing two-thirds of its share. Other notable European producers include Denmark, France and Italy.

A quantitative estimate of the global biogas production by source is given in table 10.2.

Table 10.2. Estimate of global production of biogas (Mtoe) by source (data source: [10]).

	Crops	Animal manure	MSW	Wastewater	Total
EU	8	6	3	1	18
China	<1	5	2	<1	7
USA	<1	<1	3	<1	3
Rest of the world					4
Global total					**32**

As seen in the table, Europe is the largest producer of biogas, with Germany producing two-thirds of the total biogas. The Chinese biogas programme is dedicated to household scale digester (i.e. micro-scale digester) production, which accounts for two-thirds of the installed biogas capacity of the country.

In the USA, landfills (which rely on MSW for biogas generation) provide 90% of the nation's biogas. Fifty percent of the remaining 10% of global biogas production comes from Asia and South America, with Thailand, India, Argentina and Brazil leading the way.

10.3.1.3 Uses of biogas

Biogas is used either in its original form or upgraded to **biomethane**. The latter is also known as **renewable natural gas (RNG)**, as its composition is very similar to that of natural gas.

The main uses consist of:

- Power generation, consisting of dedicated electrical power generation, combined heat and power (CHP), and combined heating, cooling and power (trigeneration) (see case study on Biowerk Hamburg chapter 6, section 6.5.4);

- Cooking (especially in the rural/remote areas of developing countries);
- Transportation (when upgraded to biomethane [12]).

In 2018, most of the biogas produced globally was used for power generation (31.4 %) or co-generation (32.7%), with heating and cooking accounting for 27.3% and the remaining 8.6.% used for upgrades to biomethane [10].

10.3.2 Biomethane or renewable natural gas (RNG)

Biogas can be used with minor cleaning/scrubbing for power generation and cooking/heating purposes. For use as an alternative transportation fuel however, it needs upgrading to RNG to meet the specifications of spark-ignition engines [12]. The unsuitability of raw biogas as a transportation fuel stems from its energy content (LHV) and impurity content. The LHV of the upgraded RNG is 36 MJ m^{-3} [10], while that of raw biogas varies between 16 and 28 MJ m^{-3}. This presents a wide range of values that is difficult to standardise, and its modal value also falls well below that of natural gas, the conventional fuel for gas-driven vehicles.

In addition, the impurities contained in biogas can be corrosive, causing potential harm to the engine. The production of biomethane is either through the upgrading of biogas (accounting for 90% of the total production) or gasification of solid biomass, followed by methanation.

10.3.2.1 Cleaning and upgrading of biogas

Biogas is converted to biomethane (RNG) via two stages: purification and upgrading (see IRENA [13], also IEA Bioenergy Task 37 [14] and the EPA report on Renewable Natural Gas [15]).

Purification involves the removal of contaminants and impurities such as water vapour and hydrogen sulphide, as well as toxic compounds such at siloxanes and volatile organic compounds (VOCs). Upgrading generally refers to the removal of CO_2 and other gases to bring the energy specifications of the resulting renewable natural gas close to that of natural gas (NG) [13].

10.3.2.2 Case study of biogas and biomethane production and use—the USA

In the USA, biogas is produced from municipal solid waste (MSW), sewage sludge, yard and crop waste, food and food-processing wastes and animal manure [15]. The essential steps to convert biogas to RNG are:

- Primary treatment
  - Removal of moisture and particulate matter from raw biogas.
- Secondary treatment
  - Additional removal of moisture;
  - Removal of siloxane, sulphur, hydrogen sulphide;
  - Compression of the biogas.
- Advanced treatment
  - Removal of CO_2, O_2, N_2, and VOCs;
  - Further compression of the gas.

In the case where the biogas is derived from landfill gas (LFG) the steps may be summarised as in table 10.3.

Table 10.3. Treatment of landfill gas (LFG) for the production of clean biogas and RNG (data source: US EPA [15]).

Level of treatment	Treatment process	Final product and use
Primary treatment	Removal of moisture and particulates from raw biogas	'Medium-Btu biogas' (i.e. medium energy content biogas), suitable for some heating applications such as leachate boilers and kilns
Secondary treatment	• Further moisture removal • Siloxane and H_2S removal • Compression	Medium-Btu biogas for heating and power generation
Advanced treatment	• CO_2, O_2, N_2, VOC removal • Further compression	• RNG for transportation (as an alternative to natural gas) • Injection into natural gas pipelines • Availability at road vehicle filling stations • Power generation, heating/cooling, hot water (*Note:* RNG is used as alternative to compressed natural gas (CNG) or liquefied natural gas (LNG)).

10.3.2.3 Carbon dioxide removal technologies

The four technologies for carbon dioxide removal from biogas that are available in the USA and other nations consist of **membrane systems**, **pressure swing adsorption**, **solvent scrubbing** and **water scrubbing**. These may be described briefly as follows:

Membrane systems use sieve action to separate molecules in a mixture according to their sizes, retaining particles larger than the pore size and transmitting the smaller particles.

In **pressure swing adsorption (PSA) systems**, the biogas is introduced under pressure to an adsorbing medium. While methane is allowed to pass through the product gas, carbon dioxide and nitrogen are retained by the medium. After the medium is saturated with the adsorbed gases, it is de-pressurised to release the adsorbed gases and is thus re-generated for further use.

Solvent scrubbing systems use a solvent (an amine or physical solvent) to remove carbon dioxide and hydrogen sulphide from the biogas stream, while allowing methane to pass through.

Water scrubbing is a simple process in which the biogas is fed into water under pressure, causing carbon dioxide to be dissolved while methane passes through. De-pressurisation at a later stage releases the carbon dioxide. The efficiency of the process can sometimes exceed 99%!

10.3.3 Future supply outlook for biogas and biomethane

As shown in table 6.2, biogas contributes only 0.08% to the total global final energy consumption. It was also mentioned in chapter 6 that only about 2% of the available feedstock for biogas is used for its production, indicating that a huge proportion of the biogas production potential was unutilised. Is it possible to increase the share of biogas in the final energy consumption in the future?

The answer will clearly depend on the factors that determine biogas production, both in the present as well as the future. So what is the future outlook for biogas (and biomethane)?

An outlook for biogas and biomethane supply and demand up to 2040 has been prepared by IEA in its World Energy Outlook Special Report for 2020 [16]. It carries out assessments based on two possible scenarios for the future: the Stated Policy Scenario (STEPS) and the Sustainable Development Scenario (SDS).

The first includes the implications of climate change policies (in particular the realisation of the nationally determined contributions by countries) and bases its assessment on the present demand for biogas/biomethane and the employment of the latest available technology for biogas production. In the SDS scenario, the assessment starts with the assumption that the aims of the sustainable development goals (and in particular SDG7—energy access for all by 2030) has been attained, and the relevant technologies are in place for this end result.

The outcomes of the two scenarios are, however, widely divergent. This reveals that the future outlook for biogas production is strongly dependent on the broader context of the global energy system. Any change in the state of this system is likely to bring about a corresponding change in the outcomes.

(See research question 10.1—Predicting the future of biogas and biomethane.)

An emerging development that will bring about landmark changes in the global energy system is the gaining importance of EVs in the transportation sector. Movement towards clean transportation in the form of zero emission vehicles was given a boost in the run-up to COP26 (November 2021) at the global policy level and has been picking up momentum at the national scale ever since.

One example is the pledge that the world's six largest automakers and 30 countries signed to phase out gasoline car sales by 2040, as reported in the New York times report of November 2021 (see the Greenwire report [17]). The **Glasgow Declaration on Zero Emissions Cars and Vans** [18] was signed on the same day as the COP26 declaration on transportation.

An example of the zero emissions commitments undertaken at the national scale is provided by the State of Victoria in Australia. Under its Zero Emissions Vehicle Roadmap, the Victorian government has legislated that by 2030, half of all light vehicle sales in Victoria will be zero emissions [19].

(See exercise 10.1—Assessing the potential impact of the switch to EVs on the global production of biomethane.)

Exercises and questions

Exercise 10.1. Assessing the potential impact of the switch to EVs on the global production of biomethane.

a) Assess the future importance of biomethane-powered road transportation vehicles on the assumption that the numbers of EVs remained the same as the present. Use 2040 as the target year.
b) Use a published estimate of the number of EVs in 2040 under their current development trend, and comment on how this will impact on the use of biomethane for road transportation, and the production of biomethane generally.
c) How can governments change the resulting scenario through policy legislations?

Research question 10.1. Predicting the future of biogas and biomethane.

a) Study the IEA outlook for biogas and biomethane and test your understanding of its outcomes by answering the following questions:
    (i) What are STEPS and SDS? Describe each scenario in full, and show how they differ from each other.
    (ii) What happens to carbon emissions under the two scenarios?
    (iii) Describe in words, and with the help of diagrams, how global bioenergy (made up of solid biomass (traditional and modern use), biofuels, and biogas (direct use and upgrading)) demand changes after the implementation of the two scenarios.
    (iv) Compare the global demand for direct use of biogas in advanced economies under the SDS scenario with that in developing countries.
b) One of the reasons for the divergence between the outcomes of the two scenarios is that they both depend on assumptions. Examples are the future status of factors such as technology or societal trends, or the efficacy of policies in achieving their future goals.
   Identify the critical factors and assumptions that were used to determine the final outcomes in the two scenarios, and examine their status. Are they reasonable? How can some of these assumptions/factors be modified to close the disparity between the two outcomes? Support your argument by carrying out a detailed analysis of at least one of these factors.

References

[1] Alternative Fuels Data Center US Department of Energy; *Biodiesel Blends* [accessed 24 July 2022]. Available from https://afdc.energy.gov/fuels/biodiesel_blends.html

[2] DieselNet Technology Guide DieselNet. *Biodiesel Standards and Properties* [accessed 10 July 2022]. Available from https://dieselnet.com/tech/fuel_biodiesel_std.php

[3] Sani Y M, Daud W M and Aziz A R 2012 Biodiesel feedstock and production technologies: successes, challenges and prospects ed Z Fang *Biodiesel – Feedstocks, Production and Applications* (London: IntechOpen) ch 4 [accessed 21 October 2022]. Available from https://intechopen.com/predownload/41600

[4] Elgharbawy A S, Sadik W A, Sadek O M and Kasaby M A 2021 A review on biodiesel feedstocks and production technologies *J. Chil. Chem. Soc.* **66** 5098–109
[5] SRS Engineering Biodiesel: SRS International; *High FFA Pretreatment* [accessed 7 August 2022]. Available from http://srsbiodiesel.com/technologies/esterification/high-ffa-pretreatment/
[6] Chai M, Tu Q, Young T Y and Lu M 2014 Esterification pretreatment of free fatty acid in biodiesel production, from laboratory to industry *US Environmental Protection Agency Papers* 212 [accessed 7 August 2022]. Available from https://digitalcommons.unl.edu/cgi/viewcontent.cgi?article=1212&context=usepapapers
[7] Thangaraj B, Solomon P R, Muniyandi B, Ranganathan and Lin L 2018 Catalysis in biodiesel production. A review *Clean Energy* **3** 2–23 Available from https://academic.oup.com/ce/article/3/1/2/5250541
[8] Deshpande S R, Sunol A K and Philippidis G 2017 Status and prospects of supercritical alcohol transesterification for biodiesel production *WIREs Energy Environ* [accessed 21 October 2022] e252. Available from https://usf.edu/pcgs/documents/wene252.pdf
[9] Saka S and Kushdiana D 2001 Biodiesel from rapeseed oil as prepared in supercritical methanol *Fuel* **80** 225–31
[10] *An Introduction to Biogas and Biomethane* (Paris: IEA) [accessed 14 September 2022]. Available from https://iea.org/reports/outlook-for-biogas-and-biomethane-prospects-for-organic-growth/an-introduction-to-biogas-and-biomethane
[11] *Global Potential of Biogas.* World Biogas Association 2019 [accessed 25 July 2022]. Available from https://worldbiogasassociation.org/global-potential-of-biogas/ for a comprehensive list of available feedstocks
[12] *Renewable Natural Gas Production.* AFDC [accessed 21 October 2022]. Available from https://afdc.energy.gov/fuels/natural_gas_renewable.html
[13] IRENA 2017 *Biogas for Road Vehicles: Technology Brief.* (Abu Dhabi: IRENA) [accessed 8 August 2022]. Available from https://irena.org/publications/2017/Mar/Biogas-for-road-vehicles-Technology-brief
[14] *IEA Task 37 – Perspectives on Biomethane* (Paris: IEA) [accessed 21 October 2022]. Available from https://ieabioenergy.com/wp-content/uploads/2022/03/IEA_transport_T37_END_HIGH.pdf
[15] US EPA 2020 *An overview of Renewable Natural Gas from Biogas* (Washington, DC: US EPA) [accessed 8 August 2022]. Available from https://epa.gov/sites/default/files/2020-07/documents/lmop_rng_document.pdf
[16] World Energy Outlook Special Report. March 2020. *Outlook for Biogas and Biomethane: Prospects for Organic Growth* (Paris: IEA) [accessed 8 August 2022]. Available from https://iea.org/reports/outlook-for-biogas-and-biomethane-prospects-for-organic-growth
[17] *Countries, Companies Pledge to Reach 100% EVs by 2040.* Greenwire [accessed 23 October 2022]. Available from https://eenews.net/articles/countries-companies-pledge-to-reach-100-evs-by-2040/
[18] *COP26 Declaration: Zero Emissions Cars and Vans*, GOV.UK [accessed 23 October 2022]. Available from https://gov.uk/government/publications/cop26-declaration-zero-emission-cars-and-vans
[19] *Zero Emissions Vehicles* Victoria State Government [accessed 23 October 2022]. Available from https://energy.vic.gov.au/renewable-energy/zero-emissions-vehicles

Chapter 11

Emerging fuels for transport and power

This chapter introduces several fuels for transportation and power generation that are either new and at a developmental stage, or have been known for many years but have attracted fresh interest. These include hydro-treated vegetable oil (HVO), syngas, pyrolysis oil, hydrogen and the products of the Fischer–Tropsch Synthesis (FTS) process.

HVO is a green alternative to biodiesel that is produced from the same feedstocks as those used for the production of the latter fuel, but with superior fuel properties than those of biodiesel. Syngas, a mixture of hydrogen and carbon monoxide, can be used either as a fuel for gasification power plants, or as starting materials for the synthesis of other transportation fuels through the FTS process. Pyrolysis oil is the liquid product of pyrolysis, and is finding use as fuels for power plants. This fuel is able to substitute for heavy fuel oil (HFO) in any of its applications. Hydrogen is a new biofuel for powering fuel cell EVs and providing a (chemical) energy storage medium.

The chapter provides brief descriptions of the nature, production and use of these novel fuels.

11.1 Introduction

The aim of part III of the book has been to provide a detailed look at all existing and emerging biofuels for transport and power generation. The previous two chapters considered bioethanol, biodiesel and biogas, the three most important biofuels in use today. They also examined biobutanol, a liquid biofuel that has been under development for some time.

This chapter considers newer biofuels that are entering the market now. The fuels investigated in the following sections are:

- Hydro-treated vegetable oil (HVO), which is a green fuel substitute for diesel engines, which has better fuel properties than biodiesel for several reasons;

doi:10.1088/978-0-7503-3555-3ch11

- Syngas, a (not altogether new) fuel that is used for power generation in gasification power plants, as a precursor for the final products of the FTS process and the production of blue hydrogen;
- Pyrolysis oil (also called bio-oil) that is emerging as a new green fuel for power generation and sustainable transportation, as well as a possible feedstock for green hydrogen production;
- Hydrogen, the new green (and blue) fuel for hydrogen fuel cell electric vehicles (FCEVs) and for energy storage devices.

We begin the next section with a short review of HVO.

11.2 HVO—a better biofuel for diesel engines

As mentioned in chapter 4, section 4.2.2, hydro-treated vegetable oil (HVO), also known as **Hydro-processed Esters and Fatty Acids (HEFA)** or simply as **renewable diesel**, provides a better alternative to FAME biodiesel by avoiding the contaminants arising during the production process of the latter, as well as bringing its chemistry closer to that of petroleum diesel.

Petroleum diesel is a mixture of pure hydrocarbons. The main green alternative to this fossil fuel today is Fatty Acid Methyl Ester (FAME) biodiesel. Unlike the fossil fuel, however, this organic fuel substitute is oxygenated. Thus, its chemical properties differ in general from those of petroleum diesel.

An issue with the oxygen content is that it necessitates the formulation of a separate fuel standard for this new fuel (see chapter 8). Thus, an important benefit of removing the oxygen is that it makes it easier to tailor the properties of the resulting hydrocarbon fuel to meet the standard specifications of diesel engines (which were originally defined for non-oxygenated fuels). The removal process involves hydrogenation and oxygen removal, followed by further treatment. The methods were first indicated diagrammatically in figure 4.3 of chapter 4, and are treated in more detail below.

11.2.1 Production of HVO

It is possible to prepare oxygen-free biofuel from the same feedstock (oils, fats) as for FAME biodiesel. This is achieved through the processes of **hydrogenation** and **deoxygenation** that leads to the production of HVO. The production of this new fuel proceeds in two steps [1].

In **step 1** (called saturation and oxygen removal), hydrogen gas is used to (catalytically):

(i) saturate any double bonds existing in the molecular structure of the feedstock molecule (a triglyceride);
(ii) cleave the fatty acid chain from the propane backbone of the triglyceride; and
(iii) remove the oxygen from the fatty acid formed above.

The result is a fully saturated hydrocarbon with the same chain length as the fatty acid molecule. This product, however, is unable to meet certain physical requirements of the desired fuel. In particular, the freezing point of the product needs to be suppressed to sustain severe cold weather conditions. This can be achieved by changing the straight-chain structure of the hydrocarbon to produce branched-chain isomers.

In **step 2** (called isomerisation and cracking), the straight-chain structure of the hydrocarbon is re-configured to branch-chain isomers which have the desired lower freezing points (e.g. freezing points as much as 23 °C and 53 °C below the −23 °C of the starting straight-chain isomer). For further information regarding the industrial manufacture of HVO, the reader is referred to the Neste Handbook [2] or other sources.

(See exercise 11.1—Understanding the chemistry of HVO production.)

HVO is currently being produced and used in several countries. The share of HVO compared to those of ethanol and FAME biodiesel in the biofuels mix has thus far been small. In 2021, the volumes of ethanol, FAME biodiesel and HVO produced globally were 105 billion, 45 billion and 9.5 billion litres, respectively [3]. The share of HVO production has been rising steadily in recent years, and has increased from 6.5 billion litres in 2019 to the present figure of 9.5 billion litres. This has occurred despite the effects of the COVID19 pandemic in 2020 (which produced a 5% decline in global biofuel production that year but failed to decelerate the rising rate of production of HVO).

11.3 Syngas—a gas for many applications

Syngas is produced when feedstocks containing carbon and hydrogen are subjected to high temperatures and pressures in the presence of water (in the form of steam) and a controlled supply of oxygen. Chapter 3 discussed the production of syngas through the processes of pyrolysis and gasification.

Syngas appears in several contexts relating to the production processes of fuels and chemicals. In particular:

- it is the main product of gasification, and provides the fuel for gasification power plants (see chapter 3);
- it provides the starting reactants for the production of blue hydrogen (see section 11.6);
- it provides the precursor to the production of synthetic fuels via the FTS process (see next section).

11.3.1 Uses of syngas

As mentioned in chapter 3, the uses of syngas include power generation via the use of Integrated Gasification Combined Cycle Power Plants (IGCCPPs), ammonia production via the Haber Process and the production of blue hydrogen.

Blue hydrogen is the name commonly used for the hydrogen extracted from syngas produced as an intermediate product in the steam–natural gas reforming process [4, 5]. The production of blue hydrogen is treated in detail in section 11.6.

The significant role that syngas plays in the FTS process is evident in the next section.

11.4 The FTS process—synthetic fuels for conventional engines

In chapter 4 it was stated that one of the uses of syngas was to provide the feedstock for the production of synthetic fuels via the FTS process. The use of syngas in the production of synthetic fuels via FTS has a long and interesting history. The process and its history have been described by Loosedrecht *et al* [6], and deserve special attention. Box 11.1 is a detailed reference to this work.

Box 11.1. The FTS process.

The **Fischer–Tropsch Synthesis (FTS) process** is a technology used to produce synthetic hydrocarbon fuels and other chemicals from coal, natural gas and (more lately) biomass. The carbonaceous feedstock is fed into a gasifier to produce syngas. This intermediate product is then put through the FTS process to produce a range of alkanes, alkenes and some alcohol as stationary/transportation fuels for petrol and diesel engines, the aviation industry, as well as chemicals for other use.

History

The first work in FTS began as early as 1902, when it was shown that a mixture of CO and H_2 could be converted catalytically into methane gas [7]. This led to further work in the 1920s, when Fischer and Tropsch showed that the same starting material could be used to produce a mixture of higher hydrocarbons that could replace petrol and diesel engine fuels.

The first commercial FTS plants became available in 1936, and similar plants were instrumental in the production of the fuel requirements of the German and Japanese war efforts during World War II. The FTS process was further developed in the USA after the war. Large-scale production was started in South Africa in 1995 by Sasol, which used the high temperature form of the process (HTFT) to produce petrol and the low temperature form (LTFT) in the production of wax.

The oil crisis of 1970 saw Sasol expanding its HTFT technology, with several other oil companies joining the production of synthetic petrol and diesel soon afterwards.

Nomenclature

The nomenclature used to describe the various possible processes for the conversion of carbonaceous sources via FTS is based on the nature of the initial feedstock. Thus, the process that begins with coal as feedstock to produce a liquid is termed 'coal-to-liquid' (CTL). Similarly, a gas-to-liquid (GTL) process describes the conversion of a gas to a liquid, biomass-to-liquid (BTL) refers to the conversion of biomass feedstock to a liquid, and a process where the feedstock could be anything (anything-to-liquid) is represented as XTL.

The stages of the FTS process

The entire process in the conversion of initial feedstock into the final product may be divided into three stages. In the first stage, the carbon source (coal, natural gas or biomass) is converted into syngas. This is then catalytically converted into hydrocarbons with a range of chain-lengths as well as oxygenated products through the FTS synthesis process. In the third and last stage, these compounds are hydrocracked and otherwise re-formed into the desired final products.

These stages are depicted diagrammatically in figure 11.1.

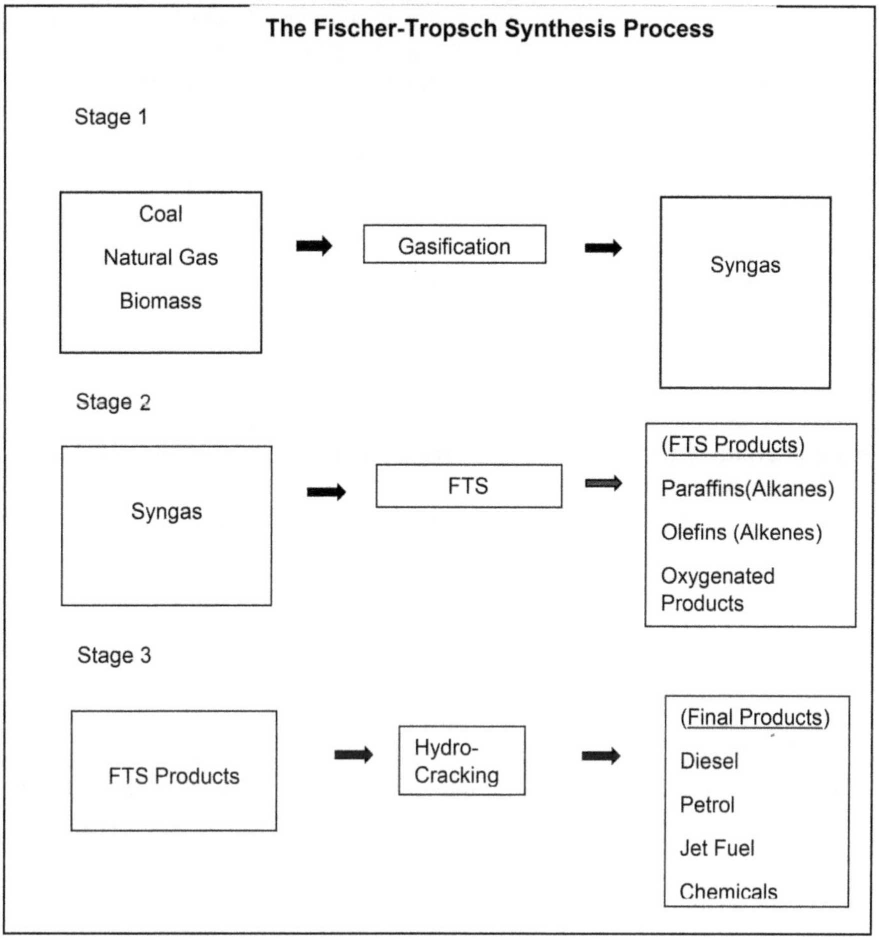

Figure 11.1. A schematic representation of the individual stages used in the conversion of the carbonaceous feedstock to the final product in the FTS processes.

11.5 Pyrolysis oil—new fuel for sea transport and power generation

In chapter 3, we learnt that pyrolysis oil is one of the three products (the others being bio-char and non-condensable or permanent gases such as CO and H_2) formed in the pyrolysis of biomass. The types of pyrolysis used today include very slow, slow, fast and flash pyrolysis, where retention times of the biomass in the reactor range from the longest (days) for very slow pyrolysis to less than seconds for flash pyrolysis. The relative yields of pyrolysis oil, as compared to bio-char and the permanent gases, increase in going from slow to fast pyrolysis (see table 3.1).

Chapter 6 provided examples of the commercial production and use of pyrolysis oil in combined heat and power (CHP) plants. At the time of writing, pyrolysis oil was being produced by a number of companies in Sweden, Finland and the Netherlands in the EU, as well as Canada, Malaysia and the USA (see table 6.2). One of the most important of these was *Btg Bioliquids* [8], which had pyrolysis oil plants in Sweden, Finland and the Netherlands. The one clearly-identifiable plant that was using pyrolysis oil for CHP was the Savon Voima plant [9] at Joensuu (see box 6.2).

The current and suggested use of pyrolysis oil extends well beyond CHP applications. According to BTG Bioliquids, bio-oil can be used:

- for heating (and CHP);
- as a sustainable transport fuel;
- for the manufacture of bio-based chemicals.

Pyrotech Energy, a producer of mobile pyrolysis plants [10], extends this list of uses to:

- fuel for burners and modified internal combustion engines;
- an alternative to fossil fuels for industrial and space heating purposes;
- bitumen production;
- biochemical production from its lignin and cellulose/hemicellulose-derived fractions;
- a fuel that provides a basis for marine fuels and aviation fuel production.

It is becoming increasingly evident that pyrolysis oil can act as a substitute for crude oil/HFO in many (if not most) of its applications. While the specific properties of the oil will depend on the feedstock from which it is derived, the utility of this fuel as an HFO substitute is evident in a comparison of the two fuels. This comparison has been carried out by Pyrotech Energy, which has labelled its process as the PyroFlash process. Table 11.1 shows comparisons of some key properties of the two fuels.

Table 11.1. Comparison of the properties of pyrolysis oil produced by Pyrotech Energy with that of heavy fuel oil (HFO) (data source: [10]).

Property	Pyrolysis oil	HFO
Moisture content (wt%)	15–30	0.1
PH	2.5	7
Specific gravity	1.2	0.94
HHV (MJ kg^{-1})	17–24	40
Viscosity at 500 °C (cP)	40–100	180
Pour point (°C)	−33	−18
NO_x (g MJ^{-1})	<0.7	—
SO_x (g MJ^{-1})	0	—

It is important to note that one of the more obvious advantages of pyrolysis oil is its physical form. As a liquid, this bio-oil is much easier to handle, transport and store than the (usually solid) biomass feedstock it is derived from.

(See exercise 11.2—Pyrolysis oil as a substitute for HFO use in diesel generators and ships.)

(See research question 11.1—In search of pyrolysis oils for road transportation diesel engines.)

11.6 Hydrogen—green fuel for EVs and storage

Hydrogen is the new biofuel that is currently gaining prominence both as a fuel for fuel cell electric vehicles (FCEVs) as well as an energy storage medium. This section considers its production and use. Before proceeding further, it is appropriate to note the proliferation of terminology that exists in the literature today that attempts to define shades of differences in the environmental quality of the gas. In one system of nomenclature [11], hydrogen is defined under a colour coding system that ranks the emissions intensity of hydrogen from green (zero emission) to brown and black (high emission). According to the Australian Conservation Foundation, this colour coding system ranks hydrogen as follows:

- **Green hydrogen** is hydrogen produced via the electrolysis of water using renewable energy as the source of electricity, and essentially has a zero carbon footprint.
- **Blue hydrogen** is hydrogen formed via steam–methane reforming followed by carbon capture and storage (CCS).
- **Grey hydrogen** is hydrogen produced from steam–methane reforming without carbon capture and storage, and has carbon emissions associated with it.
- **Brown and Black hydrogen** are types of hydrogen formed from brown coal and black coal via gasification, and have the highest amounts of carbon emissions associated with them [12].

Other colour coding systems have been proposed and are in use. For the purposes of our discussions, we will adhere to the three categories of hydrogen viz: green hydrogen, blue hydrogen with CCS and blue hydrogen without CCS.

11.6.1 Hydrogen production

The US Department of Energy—Alternative Fuels Data Center (AFDC) lists several methods used today to produce hydrogen [4].

These include:

- natural gas reforming/gasification and partial oxidation with a water–gas shift reaction;
- electrolysis of water (producing green hydrogen when renewable energy is used as the electricity source);
- renewable liquid (e.g. ethanol) reforming;
- fermentation.

It also lists high-temperature water splitting, photobiological water splitting and photoelectrochemical water splitting as emerging technologies.

11.6.1.1 Hydrogen from natural gas reforming

This process utilises natural gas (which is mostly methane) and other fuels to produce hydrogen through two processes. They are called natural gas reforming or partial oxidation [13]. The first starts with methane or other fuels and uses the **steam–methane reforming reaction** to produce hydrogen followed by the **water–gas shift reaction** to produce more hydrogen. The second process starts with the partial oxidation reaction, and then follows the same sequence of steps as the first.

The steam–methane reforming process has three steps (two reactions and a purification step):

Step 1: steam–methane reforming reaction.

Natural gas (which is mostly methane) is reacted with steam at high temperatures (700–1000 °C) and pressures (3–25 bar) in the presence of a catalyst to produce a mixture of carbon monoxide and hydrogen (i.e. syngas) according to the reaction:

$$CH_4 + H_2O\ (\text{steam}) \rightarrow CO + 3H_2 \quad (11.1)$$

Step 2: water–gas shift reaction.

The carbon monoxide formed in step 1 is then reacted with steam in the presence of a catalyst via the reaction:

$$CO + H_2O \rightarrow CO_2 + H_2 \quad (11.2)$$

This produces more hydrogen and converts the carbon monoxide to carbon dioxide.

Step 3: purification through pressure-swing adsorption (PSA).

In the last step, the carbon dioxide is separated from the hydrogen by adsorbing it in a medium under pressure (while the hydrogen passes through). The carbon dioxide is then released from the adsorber by reducing the pressure. This is the same process as that used in the purification of biogas to form biomethane encountered in the last chapter.

11.6.1.2 Hydrogen from partial oxidation of methane

In this process, the water in the steam–methane reforming reaction shown in equation (11.1) is replaced by a limited amount of oxygen (not sufficient to fully oxidise the carbon to carbon dioxide). This partial oxidation reaction of methane is given by

$$CH_4 + \tfrac{1}{2}O_2 \rightarrow CO + 2H_2 \quad (11.3)$$

In the next step, the carbon monoxide is used to produce more hydrogen through the water–gas shift reaction (11.2) as before.

In the USA, 95% of the hydrogen produced is through natural gas reforming [5].

11.6.2 Hydrogen as a green fuel

It has been mentioned that hydrogen can be either green hydrogen derived from the electrolysis of water, or blue hydrogen obtained from fossil fuels (by the two processes described above). Questions that should arise naturally are:

1. Is it possible to convert blue hydrogen to a form that is (effectively) emissions-free (i.e. zero-emissions hydrogen)?
2. Are we assured that green hydrogen has no emissions associated with it?

The answers can be obtained through a critical look at the production and use of each of these categories of hydrogen fuel. This is provided in box 11.2.

Box 11.2. When is hydrogen emission-free?

In chapter 4, it was stated that 'blue hydrogen' was hydrogen derived from fossil-fuel sources, and generally releases carbon dioxide into the atmosphere. It was also implied that hydrogen derived from the electrolysis of water is green hydrogen, and does not release the greenhouse gas carbon dioxide into the atmosphere. These assertions require further deliberation and clarification.

A green fuel is one that produces no (or insignificant) GHG emissions when it is used for energy generation. More precisely, a fuel is green if the amount of GHG (usually carbon dioxide) released per unit energy produced from it is less than that released when a similar amount of fossil fuel is used. This results in a net reduction in the emissions when the fossil fuel is replaced by the green fuel. **Note that the release of GHG includes the GHG released during the production as well as the use of the 'green' fuel.**

The two primary examples of carbon emissions that are of interest to us are those arising in electrical and mechanical power generation. In both cases, carbon emissions occur when the fuel is:

- produced from its source and transported to the point of use (e.g. a power station, a motor vehicle or a fuel cell); and
- transformed chemically in a process which is exothermic (i.e. the total chemical energy content of the products is less than that of the reactants).

Now energy is released from hydrogen through its oxidation, which may occur directly when hydrogen is burnt in the presence of oxygen, or indirectly when oxygen and hydrogen are introduced into a hydrogen fuel cell. In both cases, the only product formed is water (see equation (4.9)) and therefore no greenhouse gases are released. Thus, **no GHG emissions occur during the use of hydrogen as a fuel**. Neither green nor blue hydrogen emit carbon dioxide during their use as fuel.

It is possible, however, that **carbon dioxide release occurs in the production stage of the hydrogen**. A careful scrutiny reveals that this occurs in the case of all methods of hydrogen production listed at the beginning of this section except the electrolysis of water.

A little thought should also reveal that even in the case where the hydrogen is produced through the electrolysis of water, there may be some carbon emissions involved. This is due to the possible emissions associated with the production of the electricity used in the electrolysis that produces the hydrogen fuel. If the electricity is generated using a fossil-fuel power plant, carbon dioxide is released to the atmosphere when the fossil fuel is burnt.

Thus, hydrogen produced through electrolysis is only green if there is little or no emissions during the production of the electricity used in the electrolysis of water. This is largely true when the source of the electricity is renewable energy.

What is zero-emissions hydrogen?

Is it possible to make blue hydrogen (effectively) emissions-free? This will be so if a method could be found to capture the carbon dioxide emitted during the production of blue hydrogen. Several methods of carbon capture and storage/sequestration (CCS) have been suggested, or are being practised currently by several countries, including Australia. This topic will be the subject of further deliberation in the final two chapters of the book.

(See exercise 11.3—Blue hydrogen, green hydrogen, and emissions-free hydrogen.)

(See question 11.1—Identifying the sources of carbon emissions associated with blue and green hydrogen.)

11.6.3 Use of hydrogen—the hydrogen economy

Hydrogen is perhaps best known as a source of energy for fuel cell electric vehicles (FCEVs). It is also used in a wide variety of other applications, including portable electronic devices, as power sources for buildings and mini-grids, and providing the energy source in back-up power systems. Its utility stems from its ability to act as an energy storage medium. The two essential technologies that facilitate this role are the **hydrogen fuel cell** and the **electrolyser** used in the electrolysis of water.

Hydrogen can also be a substitute for natural gas (in the form of pure hydrogen or as a hydrogen–natural gas blend) as a fuel for domestic cooking and heating needs and other uses. This fuel can be distributed via dedicated hydrogen pipelines or existing natural gas pipelines [14].

Because hydrogen is a new fuel, one needs to create a whole new hydrogen economy to be able to produce and use it sustainably. This will include the construction of the physical infrastructure for its production, storage, transport and distribution, as well as the associated policies and legislations that facilitate such development.

The most important methods of industrial-scale production of blue hydrogen have been highlighted above. Green hydrogen is produced on the industrial scale from the electrolysis of water using renewable energy, and involves the use of large-scale electrolysers, renewable energy sources and hydrogen storage facilities.

11.6.3.1 The essential technology

In chapter 4 we saw that chemical energy stored in hydrogen can be converted directly into electricity using a hydrogen fuel cell. The opposite function of electrolysers in producing hydrogen using electricity was pointed out in chapter 7 (see box 7.1). The two devices together play an essential role in integrating hydrogen energy into energy systems.

The combination of an electrolyser and a fuel cell with hydrogen storage provides the core components of a **Regenerative Hydrogen Fuel Cell (RHFC)**. This is similar to a rechargeable battery, and can replace the lithium-ion rechargeable battery in grid electricity storage.

11.6.3.2 Role of hydrogen in grid energy storage

The list of technologies currently available for grid energy storage include **Pumped hydro (PSH), Compressed air, Molten salt (thermal storage), Lithium–ion battery, Flow battery, Hydrogen fuel cell and Flywheel** [15].

The most important of these is pumped hydro. Li-ion batteries are an electrochemical method of grid energy storage and are used widely for a variety of other storage needs. Pellows *et al* [16] argue that RHFCs may be better than Li-ion batteries at storing overgeneration of electricity, particularly in the case of wind energy. RHFCs have a lower roundtrip efficiency (30%) than lithium-ion batteries (75–90%), but similar energy return on investment when storing overgeneration from wind turbines.

11.6.3.3 Hydrogen storage

While hydrogen has high energy per unit mass, its energy per unit volume is very low at ambient temperatures. Thus, special techniques are required for its high energy density storage. These may be categorised into physical-based and material-based methods [17].

Physical-based methods involve lowering the volume via (high pressure) compression or (cryogenic) liquid hydrogen formation. Material-based methods involve the storage of hydrogen in another material. This may be via:

(i) adsorption on the surface of a material;
(ii) absorption into metals to form interstitial or complex metal hydrides; or
(iii) the formation of chemical hydrogen.

11.6.3.4 Transport and delivery

In any viable hydrogen economy, the fuel will need to be delivered from its point of production to the point of use through the most convenient and economic means. This may entail:

- compression or liquefaction of the gas;
- storage facilities; and
- gas pipelines.

In the USA, the transportation of hydrogen fuel is currently being undertaken via road using cryogenic liquid tanker trucks and gaseous tube trailers, and via pipelines to high demand regions (see chapter 13 for more details). More information is available from [18].

Exercises and questions

Exercises

Exercise 11.1. Understanding the chemistry of HVO production.

Read the descriptions of the two processes for HVO formation provided in the quoted references or other sources, and describe the chemical pathway that leads from the initial triglyceride molecule to the deoxygenated hydrocarbon.

You may start with a triglyceride which has the same R group for the three fatty acid chains, and which is 17 carbon atoms long and has one double bond.

Exercise 11.2. Pyrolysis oil as a substitute for HFO use in diesel generators and ships.

HFO is a fuel that is currently being used as substitute for diesel fuel in ships and diesel power generators.

Examine the properties of pyrolysis oil displayed in table 11.1, and make a general comment about the suitability of the pyrolysis oil as a substitute for HFO in these applications. In particular:

(i) What can you say about the engine performance (e.g. power output) with this substitution?
(ii) What are the possible risks of damage to the engine from this substitution?

Exercise 11.3. Blue hydrogen, green hydrogen and emissions-free hydrogen.

At what stage of the production and use of hydrogen does carbon emissions generally occur? Does it occur during the:

(i) Production of the hydrogen?
(ii) Use of the hydrogen?

Is carbon emission possible in the case of green hydrogen? Explain using an example of its production and use.

Is it possible to convert blue hydrogen (effectively) into zero-emissions hydrogen?

How (and when) is renewable energy useful in the production of green hydrogen?

Questions and research questions

Question 11.1. Identifying the sources of carbon emissions associated with blue and green hydrogen.

Consider the production of hydrogen gas through the processes of natural gas reforming/gasification, partial oxidation and/or fermentation. In at least two of these cases, show the key chemical steps involved in the process, and demonstrate that carbon dioxide is one of the products.

Next, describe the chemical process involved in the production of hydrogen through the electrolysis of water, and comment on the nature of the products formed.

Is this hydrogen necessarily green? To expand your answer, refer to the energy source used in the production and transmission of the electricity used in the electrolysis. Use a detailed example to demonstrate the possible emission of carbon dioxide during the production of the:

(i) electricity used in the electrolysis;
(ii) electrolysis equipment; and
(iii) other resources

required in the electrolysis of water.

When can the hydrogen produced be safely declared as green?

Research question 11.1. In search of pyrolysis oils for road transportation diesel engines.

It was mentioned in section 11.5 that the properties of pyrolysis oil depends on the feedstock. Thus, by a judicious choice of feedstock, a pyrolysis oil could be found whose properties match those of the diesel fuel appropriate for road transportation vehicles. C Wongkhorsub and N Chindaprasert (*Energy Power Eng.*, 2013, **5**: 350–55) have studied the performance of a Kubota ET-70 direct injection diesel engine driven by pyrolysis oil produced from tyres and plastic.

a) Examine their table 2, which compares the properties of the two pyrolysis oils with that of a diesel engine, and comment on the quality of these fuels as compared to the oil produced by Pyrotech Energy (see table 11.1). Which oils are more suitable for diesel engines for road transportation?
b) Find other reports from the literature of similar work done, and produce a report on the opportunities and constraints on the use of pyrolysis oil for road transportation vehicles.

References

[1] Pearlson M N 2011 A techno-economic and environmental assessment of hydroprocessed renewable diesel fuels *MSc Thesis* MIT p 106

[2] Neste Worldwide 2016 *Neste Renewable Diesel Handbook* (Rotterdam: Neste Propriety Publication) p 57

[3] REN21 Renewables 2022. *Global Status Report.* REN21; 2022 [accessed 25 August 2022]. Available from https://ren21.net/wp-content/uploads/2019/05/GSR2022_Full_Report.pdf

[4] Alternative Fuels Data Center US DoE. EE and RE. Hydrogen production and distribution [accessed 29 August 2022]. Available from https://afdc.energy.gov/fuels/hydrogen_production.html

[5] Hydrogen production: natural gas reforming US DOE. Office of Energy Efficiency and Renewable Energy [accessed 29 August 2022]. Available from https://energy.gov/eere/fuelcells/hydrogen-production-natural-gas-reforming

[6] Van de Loosedrecht J, Botes F G, Ciobica I M, Ferreira A, Gibson P and Moodley D J *et al* 2013 Fischer–Tropsch synthesis: catalysis and chemistry ed J Reeddijk and K Poepellmeier *Comprehensive Inorganic Chemistry II* vol 7 (Oxford: Elsevier) ch 7.20 pp 525–57

[7] Sabatier P and Senderens J B 1902 *Hebd Séances Acad. Sci.* **134** 514

[8] btgbioliquids Btgbioliquids. Technology to convert biomass residues into a bio-oil [accessed 31 August 2022]. Available from https://btg-bioliquids.com/our-technology/

[9] Savon Voima Savon Voima. For us Eastern Finns, clean power is commonplace [accessed 31 August 2022]. Available from https://savonvoima-fi.translate.goog/?_x_tr_sl=fi&_x_tr_tl=en&_x_tr_hl=en&_x_tr_pto=sc

[10] Pyrotech Energy Mobile Pyrolysis Plant. Pyrotech Energy. Pyrolysis oil [accessed 31 August 2022]. Available from https://pyrotechenergy.com/pyrolysis-oil/

[11] Australian Conservation Foundation Australian Conservation Foundation. Hydrogen explained: clean is green, blue is dirty [accessed 31 August 2022]. Available from https://acf.org.au/green-hydrogen-is-clean-but-blue-is-dirty?gclid=CjwKCAjwx7GYBhB7EiwA0d8oey5P9oKELFCnnqEO_90wsLxELVP3RwUfr_9qjDLWytExCTxAxLX25RoCUGwQAvD_

BwE&utm_medium=cpc&utm_source=google&utm_campaign=climate&utm_content=clean-hydrogen&ef_id=CjwKCAjwx7GYBhB7EiwA0d8oey5P9oKELFCnnqEO_90wsLxELVP3RwUfr_9qjDLWytExCTxAxLX25RoCUGwQAvD_BwE:G:s

[12] The Conversation The conversation. Explainer: how we make hydrogen from coal, and is it really a clean fuel? [accessed 31 August 2022]. Available from https://theconversation.com/explainer-how-do-we-make-hydrogen-from-coal-and-is-it-really-a-clean-fuel-94911

[13] Hydrogen production Natural Gas Reforming. Energy.gov [accessed 26 September 2022]. Available from https://energy.gov/eere/fuelcells/hydrogen-production-natural-gas-reforming

[14] Energy.gov. USDoE. Office of EE & RE. Hydrogen pipelines [accessed 31 August 2022]. Available from https://www.energy.gov/eere/fuelcells/hydrogen-pipelines

[15] Environmental Energy Study Institute 2019 EESI; Fact Sheet. Energy Storage [accessed 31 August 2022]. Available from https://eesi.org/papers/view/energy-storage-2019

[16] Pellows M A, Emmott C J M, Barnhart C J and Benson S M 2015 Hydrogen or batteries for grid storage? A net energy analysis *Energy Environ. Sci.* **8** 1938–52

[17] Energy.gov USDoE. Office of EE & RE. Hydrogen storage [accessed 31 August 2022]. Available from https://energy.gov/eere/fuelcells/hydrogen-storage

[18] Energy.gov USDoE. Office of EE & RE. Hydrogen delivery [accessed 31 August 2022]. Available from https://energy.gov/eere/fuelcells/hydrogen-delivery

Part IV

Energy, climate change and an integrated global policy framework for development

IOP Publishing

Bioenergy for Power Generation, Transportation and Climate Change Mitigation

Anirudh Singh

Chapter 12

Drivers and threats to global renewable energy production

While interest in renewable energy as a viable energy resource has existed for more than a century, at least 80% of the global energy supply is still provided by fossil fuels today. This is due largely to the global energy supply being determined by several stakeholders, who often have divergent interests. In recent decades, however, two global concerns have emerged as new drivers for the production of renewable energy. This chapter begins with an overview of the stakeholders involved in the global production and trade of fossil fuel energy, and shows how this world energy supply has played a central role in political conflicts in the past and present. It then reveals how new drivers, in the form of climate change mitigation and the desire for equity in energy access, have emerged in recent decades that are now facilitating the emergence of renewable energy as a mainstream player in the global energy market.

12.1 Introduction

The first three parts of this book dealt with the science and technology of energy production and use. Part I showed how energy is stored in biomass and converted into more useful forms. Part II showed how bioenergy is used to generate and store electrical power, and part III considered the production and use of liquid and gaseous biofuels in the transportation and power generation industries.

Part IV focuses on the people and institutions that use energy, and how energy plays a crucial role in global politics and the economic sectors that determine the welfare of people and institutions. It also examines the policies implemented by individual governments and regional and global organisations to regulate the production and use of energy.

Global energy production is determined by an intricate interplay between the political, economic and national interests of several players in world energy supply.

The mainstream of energy supply today is provided by fossil fuels, which provide at least 80% of the global energy consumption. Fossil fuel energy is produced by only some countries, but is needed by all. This has inevitably resulted in a thriving international trade in this essential commodity.

Up until a few decades ago, renewable energy was largely regarded as an alternative to fossil fuel, and thus had to compete with the latter in the global energy market. But new drivers for energy production have emerged in recent decades that advocate the use of renewable energy in preference to fossil fuels, and thus act strongly in favour of the former. These drivers include climate change mitigation and the creation of global economic equity through providing energy access for all.

There are many stakeholders involved in the production, promotion, marketing, trade and use of energy on the global scale. The aim of this chapter is to provide an overview of the nature and origins of these stakeholders, and to elaborate on the complex inter-relationships between the various sectors that act as both drivers and threats to renewable energy production on the global scale.

The chapter begins in the next section by identifying the principal stakeholders in global energy. They include the producers, users and exporters of energy. The amount of fossil fuel that exchanges hands in global trade is assessed. The international trade in this valuable commodity can result in global conflicts, and section 12.3 provides an overview of how such conflicts have impacted on the production and supply of energy in the past, and how they continue to do so today.

The role of fossil fuel subsidies as threats to renewable energy production is then elaborated on in the next section. The chapter ends by showing how climate change mitigation and the desire for energy access for all have emerged as new drivers for global renewable energy production.

12.2 Who are the stakeholders in global energy?

To fully understand the factors influencing the development of a viable renewable energy sector on the global scale, one must first consider the energy sector as a whole. An insight into how energy is produced and used in the world today is best determined by first enquiring into who the stakeholders are. These may be divided broadly into the **producers** and **users of energy**, and **the governments** which promote and regulate its use.

To proceed further, it is important to realise that while fossil fuels provide the largest share of the global energy mix today, not all countries of the world possess indigenous sources of this commodity. This has inevitably resulted in the development of a **global fuel trade** where a relatively small number of nations provide the energy requirements of the rest of the world.

The full list of stakeholders in the global energy supply thus includes:

- fossil fuel producers, comprising the countries and companies who produce the fuel;
- the exporters, who provide fossil fuels to countries who do not possess fossil fuel sources of their own;
- the traders (both fuel companies and the fuel-producing countries), who make the fuel available to the non-producing countries;

- the national governments, which legislate and regulate the use of energy;
- the domestic households and the commercial, industrial and the public sectors who use the product.

Fossil fuel in its primary form occurs mainly in the form of **crude oil**, **natural gas** and **coal**, with oil providing some 85% of the global energy needs. Table 12.1 shows the top ten oil-producing countries in the world in 2020.

The USA, Russia and Saudi Arabia head this list of ten leading oil producers, who in 2020 collectively produced 1848 Mt of crude oil, amounting to 42% of the global oil production.

Table 12.1. Top ten oil-producing countries in the world in 2020 (Mt = million tonnes) (data source: [1]).

	Country	Oil production (Mt)	Share of global total (%)
1	USA	742	16.7
2	Russia	560	12.6
3	Saudi Arabia	546	12.3
4	Canada	265	6.0
5	Iraq	234	5.3
6	China	192	4.3
7	UAE	189	4.3
8	Iran	146	3.3
9	Brazil	145	3.3
10	Kuwait	144	3.2
	Total		**71.3**
	World	**4439**	**100**

All countries of the world need crude oil or its refined products for their transportation, industry and other needs. Table 12.2 shows the most important importers of the world in 2020.

Table 12.2. Top ten oil-importing countries in the world in 2020 (data source: [1]).

	Country	Net oil import (Mt)	Share of global consumption
1	China	459	21.5
2	USA	292	13.7
3	India	226	10.6
4	Korea	151	7.1
5	Japan	151	7.1
6	Germany	85	4.0
7	Spain	67	3.1
8	Italy	63	3.0
9	Netherlands	61	2.9
10	Singapore	55	2.6
11	Others	525	24.6
	Total	**2135**	**100.2**

The data above reveals that while the USA and China rank amongst the top oil producers, they were also the two leading oil importers in 2020. This is due to their high domestic demand for the commodity.

It is important to remember that the list of fossil fuels includes gas and coal as well as oil. Indeed all three are of interest to us as sources of energy and producers of greenhouse gases. Also, as the global trade and distribution of these involves their export, the key data of interest is **the net export** of these commodities. Table 12.3 shows the list of net exporters of all three fossil fuels in 2020.

Table 12.3. The most important net exporters of crude oil, natural gas and coal in 2020 (Mt = million tonnes, bcm = billion cubic metres; numbers in brackets show the ranking of the country in the corresponding list) (data source: [1]).

	Country	Net exporters of oil (Mt)	Net exporters of gas (bcm)	Net exporters of coal (Mt)
1	Saudi Arabia	368 (1)		
2	Russia	260 (2)	265 (1)	189 (3)
3	Iraq	190 (3)		
4	Canada	148 (5)	51 (7)	28 (8)
5	United Arab Emirates	125 (6)		
6	Iran	106 (7)		
7	Kuwait	105 (8)		
8	Nigeria	93 (9)		
9	Kazakstan	70 (10)		25 (9)
10	Angola	67 (11)		
11	Qatar		124 (2)	
12	Norway		113 (3)	
13	Australia		95 (4)	393 (2)
14	USA		54 (5)	78 (5)
15	Turkmenistan		52 (6)	
16	Algeria		43 (8)	
17	Nigeria		29 (9)	
18	Malaysia		24 (10)	
19	Indonesia			448 (1)
20	South Africa			78 (4)
21	Columbia			71 (6)
22	Mongolia			28 (7)
23	Mozambique			10 (10)

Table 12.3 reveals that:

- **Russia is the leading exporter of natural gas**, and amongst **the top three exporters of oil and coal.**
- The USA, Australia and Canada are important **exporters of gas and coal**, but not of oil.

- The **leading exporter of oil is Saudi Arabia**, followed by Russia and Iraq.
- The world's leading exporter of coal in 2020 was Indonesia, followed by Australia and Russia.

These data will become very relevant later in the chapter when considering the politics of energy trade.

The global trade in energy is essentially free market trade, based on the principles of demand, supply and competition. Renewable energy (RE) must compete with fossil fuel on this market. As the fossil fuel industry has a well-established infrastructure, it has been (up until recently) able to produce and supply its products at a competitive cost in comparison to renewable energy. To promote the use of RE, governments had to use devices such as tax incentives and/or legislations.

An exceptional situation where fossil fuels lose their competitive advantage is during disruptions of their global supply chains. This occurs during political events/crises such as wars and international conflicts. The following section provides examples of two such crises.

(See exercise 12.1—Understanding the energy statistics of large industrial nations.)

12.3 Energy and global conflicts

How international conflicts can play a role in energy production and use can be seen in several events that occurred over the current and previous centuries. The onset of World War II during the last century saw a serious disruption in global energy supply, forcing nations to resort to new means (including synthetic fuels produced by the Fischer–Tropsch Synthesis process and gasification) for producing energy.

Two similar events that occurred after WWII were:

- the oil crisis of 1973, and
- the Russian invasion of Ukraine in 2022.

The political origins of the **oil crisis of 1973** can be traced back to the **Six-Day War** between **Israel** and the Middle-Eastern group of countries comprising **Egypt**, **Syria** and **Jordan** that was started by Israel on 5 June 1967 and ended six days later on 10 June [2]. During this war between Israel and neighbouring Arab states, Israel captured the **Gaza Strip**, the **Sinai Peninsula**, the **West Bank** and the **Golan Heights**, all strategic territories belonging to the Arab states.

When the war ended, the UN directed Israel to return the captured territories to their respective owners. Israel refused, resulting in continuing animosities. It eventually led to the **Yom Kippur War** of October 1973, in which Egypt and Syria attacked Israel on the Jewish holy day of Yom Kippur. They made rapid gains initially, forcing Israel to seek assistance from the USA. The USA complied, and Israel was able to fight back and advance well into the Arab states. The UN eventually passed Resolution 338 calling for an end to the fighting.

Ostensibly in retaliation to the assistance given to Israel by the USA, the Arabian members of the **Organization of the Petroleum Exporting Countries (OPEC)** stopped their oil export to the USA and several other countries in October 1973. This **oil embargo** caused oil prices around the world to escalate rapidly, producing an **oil crisis**. The USA had to take emergency measures, including fuel rationing and the adoption of energy efficiency measures, to cope with the crisis.

The crisis ended in the USA when the oil embargo against it was lifted in March 1974. It was a stern reminder to the world of the dependence of the global economy on fossil fuel supply, and how international conflicts could seriously impact the global economy through this dependency. The possibility of a repeat of this type of 'weaponising of energy' was to re-occur in the recent **invasion of Ukraine by Russia**.

On 24 February 2022, Russia launched an invasion of Ukraine, prompting quick condemnation by the UN and the Western world, including the EU countries. The NATO allies were quick to impose sanctions against Russia that included both economic sanctions as well as sanctions against individuals and entities. In the case of the EU, the sanctions so far (as at 6 October 2022) include eight packages [3].

As seen in table 12.3, Russia is amongst the top three net exporters of fossil fuels in the world, and is the leader in the case of natural gas. Most EU countries have been heavily reliant on this neighbour for their energy needs. In 2021, the EU imported 83% of its natural gas needs, 45% of which came from Russia [4].

In a statement soon after the Russian invasion, the Executive Director of the International Energy Agency (part of the Paris-based OECD) stated [5]:

'Nobody is under any illusions anymore. Russia's use of its natural gas resources as an economic and political weapon show Europe needs to act quickly to be ready to face considerable uncertainty over Russian gas supplies next winter.'

On 7 March 2022, the IEA submitted a ten-point plan to the European Council to reduce Europe's reliance on Russian gas. At the Versailles Declaration in the same month, the leaders of the region's 27 states agreed to phase out Europe's dependence on Russian fossil fuels. By the end of May the same year, the EU Council had agreed to ban 90% of all Russian oil imports by the end of the year (2022). The leaders also called for:

- further diversifying energy supply sources and routes;
- accelerating the deployment of renewables;
- further improving energy efficiency;
- improving interconnections of gas and electricity networks [6].

A complete list of EU responses to the Russian invasion of Ukraine is available at [7].

It is interesting to note that one of the actions suggested as part of the Versailles Declaration was the acceleration of the deployment of renewables. This is perhaps the first time that such a solution has been explicitly suggested as a solution to an energy crisis, and reflects the growing confidence of the world in this form of energy.

(See research question 12.1—The geopolitics behind the 1973 oil crisis.)

12.4 The role of fossil fuel subsidies

We have seen above how political conflicts can act as threats against the global supply of fossil fuels, which currently provides the mainstream supply of global energy. While this does not affect renewable energy *per se* (as long as it stays out of the mainstream), fossil fuel subsidies do.

These subsidies act as direct threats to renewable energy as they promote fossil fuels in favour of the former. Such subsidies have been defined by Oil Change International as [8]

> 'any government action that lowers the cost of fossil fuel energy production, raises the price received by energy producers, or lowers the price paid by energy consumers. Essentially, it is anything that rigs the game in favour of fossil fuels compared to other energy sources.'

As a country example, in the case of the USA, the Environmental and Energy Study Institute (EESI) [9] states that numerous energy subsidies are present in the US tax system that

> 'either promote or subsidise the production of cheap and abundant fossil fuel energy.'

In simple terms, fossil fuel subsidies are financial support offered to fossil fuel producing companies of a country to assist them in succeeding in their business enterprises. The support may range from a reduction in taxes for the oil companies, to research and development in fossil fuels, and to financial support for high-risk aspects of the enterprise such as exploration. The net result is the production of more fossil fuel (with its concomitant environmental issues) for the energy market.

The US government grants several types of tax subsidies to its fossil fuel companies and the industry as a whole to lower the cost of production and encourage domestic production of energy. These subsidies occur as **tax subsidies to the fossil fuel industry**, **support for fossil fuel research, development and deployment**, and **financing fossil fuel projects abroad**.

The tax subsidies to the fossil fuel industry may be:

- **direct subsidies** (tax incentives to companies to support and reward domestic fossil fuel production), or
- **indirect subsidies** (e.g. funding support for leasing federal land for fossil fuel extraction).

Fossil fuel research, development and deployment is funded through the US Department of Energy, which has three initiatives for funding fossil fuel research and development: The **Office of Advanced Fossil Energy R&D**, the **Loan Guarantee Program** and the **National Energy Technology Laboratory**.

Financing (US-owned) fossil fuel projects abroad occurs through the **Overseas Private Investment Corporation (OPIC)**, which is the US government's development finance institution that provides its overseas investors with financing, political risk assurance and private equity support.

A notable feature of the US government subsidies is that several of these subsidies have been around for about a hundred years, although the circumstances that prompted the introduction of the subsidies no longer exists.

To obtain an idea of the scale of these subsidies, in the case of the USA, recent (2019) estimates of direct subsidies to the fossil fuel industry are around $20 billion per year, 20% of which are allocated to coal and 80% to natural gas and crude oil. European Union subsidies are estimated to total 55 billion Euros annually [10]. According to the same source, countries offering the largest fossil fuel subsidies are China ($1.4 trillion in 2015), the USA ($649 billion) and Russia ($551 billion) per year.

The new Biden administration in the USA brought about a complete reversal of the country's attitudes and policies relating to climate change when the new president signed an executive order, on 20 January 2021, for the **USA to rejoin the Paris Agreement**. This was effected on 19 February of the same year. The new administration moved swiftly to make amends with regard to its climate change policies by signing Executive Order 14008, ***Tackling the Climate Crisis at Home and Abroad***. Amongst other things, this order repealed 13 fossil fuel tax preferences. Section 209 of the Order assured that no Federal funding was subsidising fossil fuels [11].

Subsidies for industry are not uncommon, and governments routinely offer them as assistance in the development of their industry and commercial sectors. What is significant to note, however, is that according to the International Monetary Fund (IMF), fossil fuel subsidies account for up to 85% of total global subsidies, and eliminating these subsidies would reduce global emissions by 28% [11].

Fossil fuel subsidies are not necessarily bad in all situations, and there are some instances where they are justified. An obvious example of this is improving the accessibility of fuel to remote locations (including the small Pacific Island Countries) which have to contend with supply chain issues which increase fuel prices to levels beyond affordability by the time the fuel reaches them. However, in most situations their negative environmental impacts far outweigh any perceived benefits. There has therefore been organised opposition from both activist groups (such as Oil Change International) as well as multi-national groups alike to such provisions.

Perhaps the most notable amongst the multi-national groups is the **Organization for Economic Co-operation and Development (OECD)** [12]. The aim of this 37-member international organisation is to 'build better policies for better lives … to shape policies that foster prosperity, equality, opportunity and well-being for all.'

An example of a recently-formed multi-national initiative against fossil fuel subsidies is the **Agreement on Climate Change, Trade and Sustainability (ATTS)** [13]. Spearheaded by New Zealand, this group, consisting of Costa Rica, Fiji, Norway, Switzerland and New Zealand was formed in September 2019 with the aims of:

- removing tariffs on environmental goods and new and binding commitments for environmental services;
- eliminating harmful fossil fuel subsidies;
- eco-labelling of goods.

(See question 12.1—The US position on climate change and fossil fuel subsidies.)

12.5 New drivers for global RE production

Fossil fuel is a strong competitor to renewable energy (RE) and has thus far managed to remain the mainstream provider of energy for the whole world. But the need for climate change mitigation, and the desire to bring about global equity between the rich and the poor, have renewed interest in RE as an alternative to fossil fuel.

12.5.1 Climate change

The perspective on energy changed when concern for climate change and its impacts began growing towards the end of the last century. This global crisis could not be ignored, as it was becoming clear that it was an **existential threat to humankind**. Ways and means had to be found to mitigate its impacts and to contain them within acceptable levels.

The unnatural change in the Earth's climate was being brought about by its warming, and scientific evidence pointed to a figure of merit of no more than 2 °C (and closer to 1.5 °C) as the permissible limit. What was concerning was that most of the warming was caused by greenhouse gases emitted by the use of fossil fuels. Methods had to be found for reducing these emissions, and the most obvious solution lay in replacing their use by energy sources which produced less or zero emissions.

The limelight fell on renewable energy (RE) as the best possible contender for this energy source. It quickly gained precedence over fossil fuels as the preferred energy provider, as it satisfied the specified emissions requirements. As a consequence, the mitigation of climate change has become the most powerful driver for RE production. Indeed, RE has been elevated to a status where it could now conceivably displace fossil fuels as the mainstream energy provider sometime in the future.

The next chapter provides a detailed introduction to the science of climate change, and demonstrates why it is essential that RE replaces fossil fuel to mitigate this global crisis. It also describes the energy transition that is already happening to facilitate this change in energy regime from fossil fuels to renewables.

12.5.2 Energy access for all

Another source of interest in renewable energy stems from the desire to close the gap between the world's rich and poor. Energy is needed for the sustainable development of communities as it forms one of the basic requirements for economic sustenance and growth. The world's economic infrastructure, comprising the industrial, commercial, communication and other sectors, all need energy for their operation and maintenance.

It is also known that while not all of the world's nations possess fossil fuel resources, they usually do possess renewable energy resources such as wind and solar energy. This is particularly true in the case of many small island developing states (SIDS). It is therefore possible to use such renewable resources for the provision of both grid power to urban areas as well as stand-alone power (also called distributed renewable energy) at the household level to the rural and remote areas of developing states.

Chapter 14 describes how the nations of the world have taken active steps (from as early as the year 2000) at the policy level in the sustainable development of the world's **communities.** In particular, goal 7 of the **Sustainable Development Goals** ensures that energy access to all is achieved by the year 2030.

Exercises and questions

Exercise 12.1. Understanding the energy statistics of large industrial nations.

From an examination of table 12.3 (Net Exporters), determine the following:

a) Which countries are self-subsistent in their fossil fuel needs?

b) Which countries need to import oil for their domestic consumption? Verify your answers by cross-referencing with table 12.2 (oil importers).

c) China is not shown as a net exporter in table 12.1. Does it produce any fossil fuel of its own? Refer to [1] to confirm your answer, and explain what is happening in that country. In what respects is China similar to the USA?

Question 12.1. The US position on climate change and fossil fuel subsidies.

A radical change took place in the US policies on climate change with the signing of Executive Order 14008 by President Biden in January 2021.

Provide a background to the change in the US policy on climate change that took place with the election of President Biden, and produce a brief summary of the Order that he signed. Which sections of the order are most relevant to fossil fuel subsidies?

Research question 12.1. The geopolitics behind the 1973 oil crisis.

a) The Middle-East conflict that included the Six-Day and the Yom Kippur Wars took place between Middle-Eastern neighbours, and involved fossil fuel supply as an integral part. To obtain a better insight into the events, it is instructive to gain a clearer picture of the geography of the region and the locations of fossil fuel sources within the region.

(i) Obtain a map of the region, and identify the locations of the following countries and places:

Israel, Syria, Jordan, Egypt, the Sinai Peninsula, Gaza Strip, West Bank, Golan Heights and the Suez Canal.

(ii) Create a matrix showing the fossil fuel-producing countries of the region and the fossil fuels they produce.

b) (i) What is OPEC? Study the history of this organisation, and explain which of its members were instrumental in instituting the oil embargo.

(ii) Following the oil crisis, the OECD reviewed and expanded its objectives. One of these objectives was related to the future global supply of energy. Explain what the OECD is, how it originated and what transformations it has undergone since its inception in 1948. What is the role of the International Energy Agency which was formed within its framework in 1974?

References

[1] Key World Energy Statistics 2020 IEA [accessed 31 October 2022]. Available from https://iea.org/reports/key-world-energy-statistics-2020

[2] Arab oil embargo Britannica [accessed 29 October 2022]. Available from https://britannica.com/event/Arab-oil-embargo

[3] EU response to Russia's invasion of Ukraine Council of the European Union [accessed 27 October 2022]. Available from https://consilium.europa.eu/en/policies/eu-response-ukraine-invasion/#how

[4] Infographic – where does the EU's gas come from? European Council [accessed 29 October 2022]. Available from https://consilium.europa.eu/en/infographics/eu-gas-supply/

[5] IEA unveils ten-point plan to reduce EU's reliance on Russian natural gas. Dimitris Mavrokefalidis. Energy Live News. Friday 4 March 2022. Available at https://www.energylivenews.com/2022/03/04/iea-unveils-ten-point-plan-to-reduce-eus-reliance-on-russian-natural-gas/

[6] Impact of Russia's invasion of Ukraine on the markets: EU response European Council [accessed 29 October 2022]. Available from https://consilium.europa.eu/en/policies/eu-response-ukraine-invasion/impact-of-russia-s-invasion-of-ukraine-on-the-markets-eu-response/

[7] Timeline – EU response to Russia's invasion of Ukraine. European Council. [accessed 29 October 2022]. Available from https://consilium.europa.eu/en/policies/eu-response-ukraine-invasion/timeline-eu-response-ukraine-invasion/

[8] Fossil fuel subsidies overview Oil Change International [accessed 30 October 2022]. Available from https://priceofoil.org/fossil-fuel-subsidies/

[9] Fact Sheet | Fossil Fuel Subsidies: A Closer Look at Tax Breaks and Societal Costs 2019 EESI; 29 July 2019 [accessed 30 October 2022]. Available from https://eesi.org/papers/view/fact-sheet-fossil-fuel-subsidies-a-closer-look-at-tax-breaks-and-societal-costs

[10] Ideas, Insights, Sustainable Solutions EESI [accessed 30 October 2022]. Available from www.eesi.org/about

[11] Fact Sheet | Proposals to Reduce Fossil Fuel Subsidies 2021 EESI [accessed 30 October 2022]. Available from https://eesi.org/papers/view/fact-sheet-proposals-to-reduce-fossil-fuel-subsidies-2021

[12] OECD Better Policies for Better Lives. OECD [accessed 31 October 2022]. Available from www.oecd.org

[13] Agreement on Climate Change, Trade and Sustainability (ACCTS) Negotiation New Zealand Foreign Affairs and Trade [accessed 30 October 2022]. Available from https://mfat.govt.nz/en/trade/free-trade-agreements/trade-and-climate/agreement-on-climate-change-trade-and-sustainability-accts-negotiations/

IOP Publishing

Bioenergy for Power Generation, Transportation and Climate Change Mitigation

Anirudh Singh

Chapter 13

Climate change and future energy technology

Energy is the largest contributor to climate change, and the most obvious way in which climate change can be mitigated is by changing the way we produce and use energy. This chapter shows how climate change policy, as embodied in the Paris Agreement, has led to a global energy transition in which new fuels and technologies are replacing the old in a bid to contain global temperature rise to 1.5 °C. The chapter begins by examining the goals of the Paris Agreement, and introducing the basic science of global warming. It next focuses on the goals of COP26, and considers a range of methods by which energy production and use could be changed in the transportation and power generation sectors to facilitate the achievement of net zero by 2050. It demonstrates how the countries of the world are responding to the net zero challenge by considering the case of Australia.

13.1 Introduction

This chapter aims to demonstrate how climate change policy imperatives are driving a global change in energy technology and infrastructure today.

The Paris Agreement, which laid down the rules to limit global temperature rise, was reached in 2015. However, it quickly became evident that the NDC pledges made by the Parties would be insufficient in meeting the desired goal. These pledges had to be increased significantly, and new technologies and methodologies found that assisted in the process.

The following sections show how the Paris Agreement has initiated a global change which is transforming the way the world produces and uses energy.

- Section 13.2 begins by showing what the Paris Agreement set out to do, and clearly demonstrating that these goals were not being achieved.

doi:10.1088/978-0-7503-3555-3ch13

- Section 13.3 provides a brief and succinct introduction to the science of climate change—how GHGs give rise to global warming, and how the amounts of these gases are estimated by climate scientists.
- Section 13.4 examines the goals of COP26, and provides an overview of new and emerging methods for reducing carbon emissions within the transport and power generation sectors. Most significantly, it points out the need for a transformation in the way the world generates and uses energy.
- Section 13.5 shows how Australia is meeting its net zero challenge, and how Australian universities are contributing to the effort.
- Section 13.6 is a brief look at COP27, which followed COP26 in Egypt in 2022, and notes its largely disappointing outcomes.

13.2 The net zero by 2050 ultimatum from the UNFCCC

The world knew about the threat posed by climate change due to the accumulation of human-made (anthropogenic) greenhouse gases (GHGs) in the atmosphere since well before the establishment of the UNFCCC in 1992 [1]. But it took the UNFCCC twenty years to arrive at the Paris Agreement that committed its 196 members to a common course of action to restrict global warming to an acceptable level. And it had to wait another six years of further actions and decision-making before its goals (reflected in those adopted by COP26) could be considered 'climate-ready' (or nearly so).

This section overviews the science, decisions and policy-making that linked the earliest actions by the UNFCCC to the current objectives (as embodied in the objectives and outcomes of COP26) to find a solution to the impending existential threat to humans due to climate change. It begins with the key aims of the Paris Agreement, highlights the crucial science, actions and decisions that have progressed the fight against climate change, and ends with a timeline of the COP meetings that indicates how the corresponding progress in the policy-making took place.

13.2.1 The Paris Agreement and its goals

The key goal of this agreement between the Parties of the UNFCCC, reached at the 21st Conference of the Parties (COP21) in Paris on 12 December 2015, was contained in Article 2 of the agreement. This was 'holding the increase in global average temperature to well below 2 °C above pre-industrial levels and pursuing efforts to limit the temperature increase to 1.5 °C above pre-industrial levels' [2].

Article 3 of the agreement states that the above goal is to be accomplished through the efforts of all Parties, who were to undertake successive and ambitious **Nationally Determined Contributions (NDCs)** to GHG emissions reductions in their own countries.

Article 4 of the agreement makes it clear that the cumulative NDCs of the Parties were to ensure that global GHG emissions, which were rising, were to 'reach global peaking … as soon as possible … so as to achieve a balance between anthropogenic emissions by sources and removal by sinks of greenhouse gases in **the second half of this century** … ' [2].

Since the agreement, the Parties have, according to Article 4, paragraph 12 of the agreement, been duly making NDC pledges, which have been recorded in the public registry maintained by the UN Secretariat [3].

Within a few years of the enactment of the Paris Agreement, it became apparent that global warming was not going to be contained within the required limits with the help of the existing NDCs at the time, and that greater and more urgent efforts were required by all Parties if the key goal of the Paris Agreement was to be achieved. These actions became part of the goals of COP26, and Parties were asked to update their pledges just prior to the COP26 meeting.

(See question 13.1—Understanding the goal of the Paris Agreement.)

13.2.2 The IPCC Special Report on 1.5 °C

At COP21, the UNFCCC invited the IPCC to 'provide a special report in 2018 on the impacts of global warming of 1.5 °C above pre-industrial levels and related global greenhouse gas emission pathways' [4]. This special report of the IPCC (dubbed SR15) was to be produced in 2018. The IPCC Panel accepted the invitation at its 43rd session in Nairobi in April 2016, and produced the final report on 8 October 2018.

The key outcome of this report has been quoted by the SDG Knowledge Hub [5] as follows:

> 'to be consistent with global emission pathways with no or limited overshoot of the 1.5 °C goal, global net anthropogenic CO_2 emissions need to decline by about 45 per cent from the 2010 level by 2030, reaching net zero around 2050.'

This requirement in emissions reduction was far in excess of the anticipated reductions at the time, and pointed clearly to the need for a drastic revision of the NDCs. They also indicated (for the first time) a need to reach 'climate neutrality' or 'net zero' (where emissions from all sources are equal to the removals by all sinks) closer to the middle of the century rather than some time in the second half, as stipulated in the Paris Agreement.

The publication of this report was to receive significant attention at the next COP meeting (COP24 in Katowise, Poland), which was due to begin just a few weeks after its release.

13.2.3 High Level Dialogue on Energy (HLDE)

Just five weeks prior to the start of the Glasgow COP26, the UN organised a High Level Dialogue (HLDE) on Energy in New York [6]. This meeting aimed to accelerate the achievement of two separate energy-related goals—the UN

Sustainable Development goal 7 (SDG7) and UNFCCC's goal of achieving net zero emissions by 2050.

Apart from announcing a Global Roadmap for SDG7 and $400bn as energy compacts to achieve SDG7, the summary document made recommendations for 2030 that included:

- 8000 GW of renewables by 2030 (doubling the RE share in the global energy mix from 20% to 40%);
- reducing fossil fuel share to 30% of the global energy mix;
- doubling the global improvement in energy efficiency;
- the phase-out of coal within the Organization for Economic Co-operation and Development (OECD).

It predicted that with the help of the above provisions, net zero emissions would be achieved, 92% of power would come from renewable technologies, and coal would be phased out globally by 2050.

13.2.4 The Emissions Gap Report 2021

Since 2010, the United Nations Environmental Programme (UNEP) has produced an annual report that estimates the gap between the future GHG emissions if all countries implemented their mitigation pledges (via their NDCs) and what the pledges should be to avoid the worst impacts of climate change (which translates to meeting the key goal of the Paris Agreement). These reports also identify ways to bridge the emissions gap. They are co-produced by the UNEP and the UNEP Copenhagen Climate Center (UNEP-CCC).

The latest report, titled *Emissions Gap Report* 2021. *The heat is on: a world of climate promises not yet delivered,* [7] was published on 26 October 2021. Its ten findings are a cause of both concern and alarm. They show that the NDC pledges of the Parties fall far short of achieving the near-term (2030) target, and if business was allowed to proceed as usual, it would lead to a temperature rise of as much as 2.7 °C by 2050.

The key findings, as paraphrased from the list as provided in the executive summary of the document, are shown in table 13.1.

In addition to pointing out the inadequacies of the (then) current NDC pledges, this report also reveals a serious lack of commitment by some of the members of the G20 group towards meeting their NDC obligations. Its predictions are similar to those of the IPCC Special Report on 1.5 °C as well as other recent reports. The Gap Report clearly indicates that urgent action is needed to find new ways to increase NDC commitments if the key goal of the Paris Agreement is to be achieved.

The above lists just three of the several instances where the outcomes of scientific research and decision-making since the Paris Agreement has consistently shown the

Table 13.1. Key findings to the Emissions Gap Report 2021 (data source: [7]).

1.	After a drop of 5.4% due to COVID-19, CO_2 emissions are continuing to rise globally.
2	The total effect of the new NDC pledges for 2030 will have insufficient impact on global emissions.
3	The G20 group members are not on track to meet their new or original NDC pledges.
4	Long-term (2050) net zero pledges have been made by 50 Parties, but they contain large ambiguities.
5	Only a few of the G20 group targets show clearly how their emissions paths lead to net zero by 2050. There is a need to establish near-term (2030) targets and include actions that reveal a credible pathway to the 2050 target.
6	The emissions gap is large. The new pledges for 2030 produce only 7.5% reductions over the previous emissions. To attain the 2 °C target requires a 30% emissions reduction, whereas a 1.5 °C target requires a 55% reduction in emissions.
7	Global warming is expected to increase by 2.6 °C if all the conditional and unconditional pledges are implemented. If the 'net zero by 2050' pledges are also fully implemented, warming will be restricted to 2.2 °C.
8	The opportunity created by COVID-19 to move towards low carbon economies has been missed in most countries.
9	Reduction in methane emissions from the fossil fuel, waste and agricultural sectors provides a credible means of closing the emissions gap.
10	Carbon markets can also reduce emissions, but they must be administered properly.

need to review NDC targets to improve the likelihood of meeting the goal of the Agreement. How they have impacted on the accompanying policy decisions being made by the UNFCCC may be gleaned from the timeline of the COPs and their achievements [8] (see box 13.1).

Box 13.1. The timeline to COP26 (data source: [8]).

The first meeting of the UNFCCC (the first Conference of the Parties, called COP1) was held in Berlin in 1995. It took 20 years for the Conference to arrive at a formal agreement (the Paris Agreement) amongst all its Parties to act collectively on a defined goal, and another six years to forge it into a form appropriate for the actual realisation of the original goal.

The timeline below (taken largely from [8]) traces the history of the Conference of the Parties (COP) from its beginning in Berlin in 1995 to the recent meeting (COP26) that took place in Glasgow in 2021. It highlights the progress made on the policy front that was required to transform the original implementation plan of the UNFCCC into a recipe that is close in form to being workable.

Name of COP	Place and date	Achievements
COP1	Berlin, Germany, 1995	This first UNFCCC Conference of the Parties (COP1) met from 28 March to 7 April in Berlin. The participants were concerned about the current arrangements for meeting their commitments, which involved the **Body for Scientific and Technological Advice (BSTA)** and the **Subsidiary Body for Implementation (SBI)**, and agreed to implement the actions jointly on an annual basis instead.
COP3	Kyoto, Japan, December 1997	The developed (industrialised) countries met and signed the **Kyoto Protocol**, under which they agreed in principle to reduce their greenhouse gas emissions. The details were to be decided later.
COP13	Bali, Indonesia, 2007	The Bali Roadmap was created to set up a timetable for replacing the Kyoto Protocol with a new one that included all countries, **including the developing countries** which were not included before.
COP15	Copenhagen, Denmark, 2009	The first agreement to keep global warming below 2 °C is made. Also, developed countries agree to finance the developing countries in the long term.
COP18	Doha, Qatar, 2012	Decision reached by most Parties to extend the Kyoto Protocol to 2020. USA, China, Russia and Canada did not agree.
COP20	Lima, Peru, 2014	All countries agree for the first time to develop and share their commitments to reducing greenhouse gas emissions.
COP21	Paris, France, 2015	The Paris Agreement is accepted by all. It is agreed that global warming should be kept below 2 °C above pre-industrial levels, and efforts continued to limit it to 1.5 °C.
COP22	Marrakesh, Morocco, 2016	These proceedings took place in an uncertain atmosphere due to the US elections that were taking place and the possibility that Donald Trump would become President. The Marrakesh Action Proclamation announced continuing political commitment to the Paris Agreement, the Marrakesh Partnership was proclaimed to strengthen and maintain climate collaboration to 2020, and the first meeting of the CMA, the decision-making body of the Paris Agreement, took place.
COP23	Bonn, Germany, 2017	Progress was made on the **Paris Rulebook**, which sets out the rules for implementing the Paris Agreement in practice. The Talanoa Dialogue was also introduced as a method for discussions and decision-making.
COP24	Katowice, Poland, 2018	The Paris Agreement Work Programme (Rulebook) was debated and an agreement finally reached. It was set to go into operation in 2020. The just-released IPCC Special Report on 1.5 °C was noted and received with mixed support. There was debate over the wording. Greta Thunberg gave her powerful message to the global community.
COP25	Madrid, Spain, 2019	A compromise deal was agreed for curbing carbon emissions globally. The discussion on Article 6 (Carbon markets) of the agreement was debated intensely, with strong disagreement/demands from Australia, India and Brazil. This issue was postponed until COP26.
COP26	Glasgow, UK, 2021	More than 100 countries agreed to cut methane emissions and signed a pact to end deforestation. Further, 140 countries agreed to strengthen their 2030 targets and 190 countries agreed to phase-out coal power. The Paris Rulebook was finalised.

As energy is the main emitter of GHGs and thus the main driver of climate change, the physical implementation of the scientific outcomes and policy developments noted above will involve this commodity in a significant way. In particular, the world will need to transform the way it produces and uses energy. This in turn will entail a similar transformation in energy technology and infrastructure on a global scale. Before considering how this is actually happening, it is instructive to consider the scientific basis of global warming.

(See exercise 13.1—Tracing the history of the global warming targets.)

13.3 The science behind climate change

13.3.1 Global warming

13.3.1.1 The Earth's energy budget

Global warming occurs when the Earth's energy budget (the balance between incoming energy from the Sun and the outgoing radiant energy from the Earth) is disturbed. One way this can happen is through the presence of greenhouse gases in the atmosphere.

The Earth's surface receives energy from two sources—the Sun above and the interior of the Earth below. The interior source is largely geophysical in nature, including energy stored from the time of the Earth's formation, as well as nuclear and chemical sources. But it is tiny, and can be ignored for our purposes. Thus, the Earth's energy budget is determined by the incoming solar radiation (which is short wavelength radiation) and the outgoing (long wavelength) radiation from the Earth. When the two are equal, we say the system is in 'radiative equilibrium'.

Note that there are numerous other energy flows that take place close to the Earth's surface. These include wind motion and ocean circulations due to temperature differentials caused by uneven radiative power arriving from the Sun at different regions of the surface (e.g. the equator and the poles), temperature and density differentials producing conduction and convection, energy transport through evapotranspiration, etc.

All these flows contribute to the Earth's climate system, but are largely irrelevant to the energy budget as it only considers **the net incoming energy flux to the Earth from outer space** and the **net outgoing flux to outer space**. These flows are (barring the entry into the atmosphere of meteors and other massive bodies and the escape of Earth-launched space vehicles from Earth's orbit) always in the form of electromagnetic energy.

13.3.1.2 Incoming (short wavelength) solar energy

The intensity of solar radiation one Earth–Sun distance (i.e. 1 AU) away from the Sun is 1360 W m^{-2}, called the solar constant. But because the Earth's surface is a sphere, which has a surface area four times that of a circle, this value averages out to be one quarter of the solar constant over the entire surface and the year.

Thus, the average solar energy received at the top of the atmosphere (TOA) is approximately equal to 340 W m^{-2} [9, 10]. This incoming solar energy flux is either

absorbed by the Earth–atmosphere system or **reflected** away from the Earth. The reflected radiation is made up as follows:

- about 77 W m^{-2} is reflected back to space by the atmosphere and clouds;
- some 23 W m^{-2} is reflected back by the Earth's surface.

This means that about 240 W m^{-2} is absorbed by the Earth–atmosphere system.

This **Absorbed Solar Radiation** (**ASR**) may be expressed in terms of the effective reflectivity, or **Albedo** (***A***) of the Earth according to equation (13.1) as follows:

$$\text{ASR} = (1 - A) \times 340 \tag{13.1}$$

which gives the effective value of the albedo (also called the Bond Albedo) as approximately 0.3.

13.3.1.3 Outgoing (long wavelength) radiation (OLR)

Despite the complex nature of the Earth–atmosphere system that produces the outgoing long wavelength radiation (OLR), this system may be modelled as a black body radiator. This radiation is thus described in terms of the Earth's surface temperature by the Stefan–Boltzmann Law for black body radiation (see box 13.2). We can thus say that

$$\text{OLR} = \sigma T_{\text{Skin}}^4 \tag{13.2}$$

where T_{Skin}^4 is called the Earth's surface skin temperature.

Box 13.2. What is short wavelength and long wavelength radiation?

Black bodies give off a spectrum (i.e. a range) of wavelengths of light, peaking in intensity at $\lambda_{\max}$. It is found that the peak wavelength is inversely proportional to the temperature T (measured in kelvins) (i.e.

$$T\lambda_{\max} = \text{constant}(=0.2898 \text{ cm} - \text{K})$$

This inverse relationship between the peak wavelength and the temperature was discovered by Wein, and is called **Wein's Law (or sometimes Wein's Displacement Law)**. It shows that the peak wavelength shifts to smaller values as the temperature of the body increases (and vice versa).

Now consider the Sun and the Earth. The Sun's surface temperature is very high (~6000 K). Thus, the **peak wavelength of solar radiation is a small number (i.e. the wavelength is short)**. The Earth's surface temperature, on the other hand, is relatively low (~300 K). Thus, **the peak wavelength of the radiation emitted from the Earth's surface is large, or the wavelength is long**, and usually lies in the infra-red region.

The terms 'short or long wavelength' are used in a colloquial way to point to this difference in the nature of the black body radiation from the Earth and the Sun.

To complete the picture, the intensity of the radiation flux (i.e. the total energy of the radiation given out per unit time) is also related to the temperature. This relationship is given by another law called the **Stefan–Boltzmann law**. This states that the radiant power P emitted per unit surface area A of a black body is proportional to T^4; i.e.

$$P/A = \sigma T^4 \tag{13.3}$$

where σ, called the Stefan–Boltzmann constant = 5.67×10^{-8} W m^2 K^{-4} [11].

Thus, as the temperature of the body increases, the energy of the radiation emitted by it per unit area of the surface increases very rapidly.

Now the energy of electromagnetic radiation is inversely proportional to its wavelength (picture radiation as photons). Thus, the peak wavelength of the radiation shifts to smaller values (i.e. shorter wavelengths) as the energy increases.

To summarise,

- The thermal behaviour of all bodies may be approximated by that of an ideal body called a black body.
- All bodies radiate energy in the form of electromagnetic radiation **simply because they are hot** (i.e. their temperature is above 0 K).
- This radiation appears in a spectrum of wavelengths, each characterised by a peak wavelength which depends inversely on the temperature.
- The **value of the peak wavelength decreases** (i.e. shifts or displaces to smaller values) as the temperature increases, in accordance with Wein's Displacement Law.
- The **intensity of the energy given out per unit time** (i.e. the power given out) **increases very rapidly** as the temperature increases.

The use of the terms 'short wavelength' and 'long wavelength' are elaborated on in box 13.2.

(See exercise 13.2—Comparing Wein's Law and Stefan–Boltzmann's law.)

13.3.1.4 Understanding the dynamics of the energy budget

The energy budget means that at equilibrium

Outgoing longwave radiation (OLR) = Absorbed solar radiation (ASR)

The ASR is, to a good approximation, constant throughout the year. The Earth's temperature at any location is determined by this received solar flux, and will increase as the Earth turns from night to day. This increase in temperature will effect a corresponding change in the OLR, which will stop increasing when a new equilibrium is established. We know this from our common experience. (The Earth's temperature does not keep rising after the Sun has been up for some time, but reaches an equilibrium value. At this temperature, the ASR becomes equal to the OLR.)

What happens to the Earth's temperature if we introduce a new agency in the atmosphere that reduces the value of the OLR? One such agency consists of the greenhouse gases.

(See exercise 13.3—Why a GHG increases the Earth's temperature and causes global warming.)

13.3.1.5 Radiative forcing and the Earth's temperature

What is radiative forcing?

Introduction of GHGs into the atmosphere effectively reduces the amount of outgoing radiation flux, disturbing the radiative balance. To re-establish this balance, the Earth has to warm up.

The change in radiative flux ΔF is related to:

- the amount of GHG added to the atmosphere;
- how good the GHG is at reducing the radiation flux.

Calculating the radiative forcing ΔF and global warming ΔT due to a GHG:

Relationships between ΔF and the resulting ΔT involve the concentrations of the GHG and weighting factors. Expressions for the calculation of ΔF from GHG concentrations are listed by NOAA Global Monitoring Laboratory in table 1 of [12].

In the case of carbon dioxide, the expression is

$$\Delta F = \alpha \ln (C/C_o) \tag{13.4}$$

where $\alpha = 5.35$, C is the current atmospheric carbon dioxide concentration in ppm, and C_o is the concentration that existed in 1750.

To see a simple way of estimating the amount of global warming ΔT produced by the radiative forcing ΔF, see e.g. [13].

13.3.1.6 Annual Greenhouse Gas Index (AGGI)

It is instructive to know how rapidly global warming has been increasing over the years due to all GHGs. A convenient way of doing this is to first calculate the radiative forcing in any year due to the combined effect of all the GHGs. Next, this value can be compared to that in any fixed year, producing the Annual Greenhouse Gas Index (AGGI). This reference year, which has been chosen as 1990, has an AGGI of 1 [12].

The results show that in 2021, global warming had increased by 49% over its value in 1990. This is an alarming result, when one considers that it took more than 200 years for the warming to increase from its 1750 value of nil to what it was in the reference year of 1990.

Table 13.2 lists the radiative forcing, CO_2 concentration and AGGI of the main GHGs for selected years from to 2021.

Table 13.2. Radiative forcing, CO_2 concentration and Annual Greenhouse Gas Index for atmospheric GHGs for selected years (source: [12]).

	Radiative forcing (W m^{-2})								
Year	CO_2	CH_4	N_2O	CFCs	HCFCs	HFCs	Total	CO_2-eq (ppm)	AGGI
1980	1.027	0.406	0.104	0.154	0.008	0.001	1.7	386	0.808
1990	1.294	0.459	0.129	0.263	0.018	0.003	2.166	417	1.00
2000	1.515	0.481	0.151	0.281	0.032	0.008	2.468	441	1.14
2010	1.794	0.491	0.174	0.266	0.048	0.021	2.795	469	1.29
2020	2.111	0.520	0.206	0.248	0.057	0.041	3.183	504	1.47
2021	2.140	0.526	0.210	0.246	0.058	0.044	3.222	508	1.49

13.3.1.7 Greenhouse gases and global warming potential

GHGs are not equally effective in causing global warming. They firstly differ in their ability to absorb radiation. Secondly, they are not equally stable, breaking down to harmless substances over different periods of time. This is the same as saying that they have different lifetimes. The global warming potential (GWP) of a GHG is a measure of how much radiation a fixed mass will absorb over a fixed time. The GWP of a gas is defined as the amount of (long wavelength) radiation one tonne of the gas will absorb over either 100 years or 20 years, with the former definition used more commonly. In both definitions, the results are stated in terms of the GWP of carbon dioxide, which is assigned a value of unity.

Table 13.3 lists the 100-year GWPs of the most important GHGs as they were determined in the IPCC Fifth Assessment Report (AR5) [14].

Table 13.3. A list of the important GHGs and their GWPs as accepted in the IPCC AR5 (source: [14]).

Name	Chemical formula	GWP 100-year time horizon
Carbon dioxide	CO_2	1
Methane	CH_4	28
Nitrous oxide	N_2O	265
CFC-11	CCl_3F	4660
CFC-12	CCl_2F_2	10 200
CFC-13	$CClF_3$	13 900
CFC113	CCl_2FCClF_2	5820
CFC-114	$CClF_2CClF_2$	8590
CFC-115	$CClF_2CF_3$	7670
Halon-1301	$CBrF_3$	6290
Halon-1211	$CBrClF_2$	1750
Halon-2402	$CBrF_2CBrF_2$	1470
Carbon tetrachloride	CCl_4	1730
Methyl bromide	CH_3Br	2
Methyl chloroform	CH_3CCl_3	160

13.3.2 Estimating emissions

The Paris Agreement required each Party to prepare, communicate to the UNFCCC and periodically update a National Inventory of GHG emissions by sources and removals by sinks (see Paris Agreement Article 13, etc. [2]). The methodology for the preparation of such inventories are contained in the IPCC 2006 Guidelines [15] and its 2019 refinement [16].

This method of estimating emissions begins by relating all human-made (anthropogenic) emissions to human activities. The **amount of activity** is measured in terms of **activity data**, and the nature of the activity by an **emission factor**, which determines the amount of emission per unit activity. This leads to the simple relation:

Emission = Activity data × Emission Factor (EF)

In the simple example where the emission is due to fuel consumption by vehicles, the **activity is fuel consumption**, the **activity data is the amount (volume or mass) of fuel consumed**, and the **emission factor is the amount (tonnes) of CO_2 emitted per unit fuel** consumed.

The emissions and removals of GHGs in a country are envisaged as arising from five different sectors. These are:

- Energy
- Industrial Processes and Products Use (IPPU)
- Agriculture, Forestry and other land use (AFOLU)
- Waste and
- Other.

Volume 1, chapter 1 of the Guidelines gives the list of gases considered in the IPCC methodology. This includes the important emitters:

- Carbon dioxide
- Methane
- Nitrous oxide
- Hydrofluorocarbons (HFCs)
- Perfluorocarbons (PFCs)
- Sulphur hexafluoride (SF6)
- Nitrogen trifluoride

amongst others.

There are three levels or **Tiers** at which the emission estimates may be carried out. **Tier one** is the simplest and the most approximate, and **uses default values of emission factors**. **Tier 2 uses country-specific emission factors** and is more accurate than the Tier one method. **Tier 3** is the most complex and demanding, and also the most accurate of the three methods for emissions estimates.

A detailed and non-technical introduction to the use of the IPCC methodology is provided by Mani [17] for the Pacific context.

13.4 COP26 and the global energy transition

In section 13.2 it was noted that since the Paris Agreement had been enacted in 2015, a large body of scientific evidence had emerged showing that the NDC pledges (and their updates) made by the Parties were grossly inappropriate for achieving the net zero by 2050 target that was sought. To achieve this aim, significant changes had to be made to the pledges as they existed prior to COP26.

What were the necessary changes, and how could policy be brought to bear to effect these changes?

As energy is the largest emitter, one expects this to be reflected in changes to the way energy was being produced and used. Also, as carbon removal from the atmosphere occurs through carbon sinks, policy statements that safeguard and enhance such sinks are expected. How well did the goals of COP26 address these requirements, and what are some of the specific ways in which energy production and use could be tailored by the Parties to reduce net emission to zero by 2050?

This section begins with a look at the goals of COP26, and next considers the nature of the energy transition required to bring about effective reductions in GHGs.

13.4.1 Goals of COP26

The goals of COP26 were presented under four headings [18]. In brief, they were to

(a) **Secure global net zero by mid-century and keep 1.5 degrees within reach**

- accelerate the phase-out of coal;
- curtail deforestation;
- speed up the switch to electric vehicles;
- encourage investment in renewables.

(b) **Adapt to protect communities and natural habitats**

- protect and restore ecosystems;
- build defences, warning systems and resilient infrastructure and agriculture to avoid loss of homes, livelihoods and even lives.

(c) **Mobilise finance**

- mobilise **at least $100bn in climate finance per year** by 2020.

(d) **Work together to**

- finalise the Paris Rulebook (the detailed rules that make the Paris Agreement operational);
- accelerate action to tackle the climate crisis through collaboration between governments, businesses and civil society.

In summary, the four goals consider

- specific methods of reducing emissions via reducing emissions of GHGs into the atmosphere or their removal by sinks;
- adapting to the impacts of climate change through building resilience to them and rebuilding the damage to the biosphere caused by these impacts;
- facilitating action through funding support;
- adopting a collaborative approach to addressing the climate crisis.

There is a clear message for the need for urgent action. This is evident in the call for accelerated action in goals 1 and 4. In the case of energy, this may necessitate the need to resort to new methods and opportunities to meet the challenge. The next section presents an overview of what some of these solutions could be.

(See research question 13.1—Motivation for the structure and contents of the COP26 goals.)

13.4.2 New tools and opportunities for achieving net zero

Goal 1 of COP26 refers to the phasing out of coal, and the acceleration of the switch to EVs. It also advocates investment in renewables. To investigate the implementation of this goal, it is instructive to address the following questions:

- What is the role of coal-fired power plants in global emissions? Can such plants be replaced by existing and/or new zero-carbon energy technology in a seamless manner?
- What is the range of renewable energy technologies that can be considered? What are the requirements that must be satisfied for this energy transition?
- Are there additional (e.g. infrastructure) requirements for the successful utilisation of new fuels?
- More than a quarter of global emissions can be attributed (directly or indirectly) to industrial processes, in particular those encountered in the manufacturing industry. How can such commodities be made less carbon intensive? If they form part of international trade, will these new initiatives require the enactment of new trade agreements and other geopolitical arrangements?

This section attempts to answer some of these questions by carrying out a survey of the challenges and opportunities that exist for bringing about an energy transition that can successfully deliver the required outcomes. We begin with an overview of the current status of global greenhouse gas emissions and the NDC pledges of the Parties.

13.4.2.1 The current status

Table 13.4 shows the global greenhouse gas emissions for selected years since 2000 [19].

The table reveals a steady rising trend in global annual emissions except for a small dip in 2021. The global total was 50.1 Gigatonnes in 2021, which was 4.4%

Table 13.4. Global greenhouse gas emissions (in billion tonnes, i.e. Gigatonnes, of CO_2 eq) for selected years between 2000 and 2021 (data source: [19]).

Year	2000	2005	2010	2015	2016	2017	2018	2019	2020	2021
Emissions	35.84	40.57	44.88	46.87	47.53	48.25	49.37	49.76	52.4	50.1

lower than the 52.4 Gigatonnes in 2020, the difference being due to the economic impact of the COVID-19 lockdowns.

As was noted earlier, the Emissions Gap Report 2021 had found that global warming was expected to increase by 2.6 °C if all the conditional and unconditional pledges were implemented. And if the 'net zero by 2050' pledges were also fully implemented, warming would be restricted to 2.2 °C. This best-case scenario was still significantly outside both the 2 °C and the 1.5 °C requirements for the climate goal of the Paris Agreement.

These fears were later supported by the UNFCCC's **Updated Synthesis Report** [20], prepared a month prior to COP26, which confirmed the worrying trend that had been noted relating to the NDC commitments of the Parties. The UNFCCC had requested all Parties to update their NDC pledges (showing their most recent pledges for emissions reductions) prior to COP26. It then amalgamated this most recent data to produce a synthesis report. This updated report confirmed that 'for all available NDCs of all 192 Parties taken together, a sizable increase, of about 16%, in global GHG emissions in 2030 compared to 2010 is anticipated' [20].

The Intergovernmental Panel on Climate Change (*IPCC*) showed that

'such an increase, unless changed quickly, may lead to a temperature rise of about 2.7 °C by the end of the century.' (see [7]).

All this was unequivocal evidence that the emissions reductions targets had been seriously under-estimated and/or were not being adhered to, and that drastic action was required if the Paris Agreement goal was to be achieved. The emissions situation was particularly severe in the case of coal.

13.4.2.2 The challenge

Global emissions may be divided by sector into Energy (73.2%), Agriculture, Forestry and Land Use (18.4%), Direct industrial processes (5.2%) and Waste (3.2%) [21]. The sectors of immediate relevance to this book are energy and direct industrial processes, which together account for 78.4% of emissions. Power generation using coal is the greatest concern.

13.4.2.2.1 Coal

The world currently relies heavily on coal-fired power stations for its grid electricity generation. Table 13.5 shows the top ten global producers of electricity from coal [22].

As the table reveals, in 2020, China, India, the USA and Japan were the four top producers of electricity from coal. Together, they were responsible for over three-quarters (76%) of the world's coal-fired electricity generation (which stood at 6626 TWh). The country with the largest share of coal-fired generation in its national energy mix was South Africa (86% as compared to China's 61%). Australia ranked 10th in the list of coal-fired power emitters, but still generated 54% of its electricity from this source.

Emissions from coal-fired power generation is disproportionately large. In 2018, coal-fired power stations produced 10.1 Gigatonnes of emissions [23]. This

Table 13.5. Coal-fired power generation, and share of total national power mix in top 10 coal-power producing countries in 2020 (data source: [22]).

World ranking	Country	Electrical energy (TWh) generated from coal in 2020	Share of total power generation in the country
1	China	4631	61%
2	India	947	70%
3	USA	774	19%
4	Japan	274	29%
5	South Korea	192	36%
6	South Africa	191	86%
7	Indonesia	168	60%
8	Russia	155	15%
9	Vietnam	141	53%
10	Australia	135	54%

amounted to a staggering 30% of the total fossil fuel emissions (of 33.2 Gigatonnes) or 20% of the global total (of 49.4 Gigatonnes).

It is clear that coal use needs to be reduced drastically and urgently to limit the unacceptable levels of emissions arising from this single source. The two main challenges that are encountered in this endeavour are:

- Coal currently forms an indispensable part of large-scale electricity production on a global scale. Weaning the countries out of this coal dependency will be an enormous and daunting task.
- In common with oil and gas, coal-fuelled power stations are sources of despatchable power to the grid. This type of energy is needed in grids to ensure power is available at a moment's notice. Such an energy source is crucial for maintaining the baseload demands of grids, especially those incorporating intermittent power sources such as wind and solar.

Replacement of coal-fired power stations is thus not as easy as it first appears. It requires careful consideration to grid-stability issues, and will involve the incorporation of technology such as energy storage devices to successfully effect the phasing out of coal. Additionally, the transformation will most likely have a significant socio-economic impact that cannot be ignored.

13.4.2.2.2 Transportation

Despite decades of attempts at greening transportation through the introduction of liquid and gaseous biofuels as alternative fuel, transportation is still heavily dependent on fossil-fuel consuming internal combustion engines (ICE). It is currently responsible for 24% of direct CO_2 emissions from global fuel combustion [24]. While many countries have now committed themselves to transitioning to clean transportation in the form of EVs, this is by no means a universal undertaking.

An important question that remains unanswered is whether ZEVs can be introduced quickly enough to contribute their share towards achieving net zero by 2050. The establishment of the required infrastructure calls for the provision of charging stations across nations, and the development of the most appropriate battery technology. To make the whole venture a success, prices need to be brought down significantly.

Several issues must be borne in mind while progressing the development of ZEVs. Two of these are:

- Battery resources are crucial for success. But there is a global scarcity of elemental lithium for Li-ion rechargeable batteries used in ZEVs, and there is no assurance that there will be sufficient supply of such batteries for a concerted global ZEV programme.
- New sodium-ion batteries that can replace the current Li-ion for rechargeable batteries have been proposed and appear to be nearly market ready [25, 26]. But the first commercial version may appear only after 2024. Will this be early enough to meet the enhanced NDC requirements for net zero by 2050?

There is an option to resort to internal combustion engines using clean fuels, including traditional biofuels (biodiesel, ethanol), the newer HVO, and biogas/biomethane during the interim period while ZEVs are being fully developed and market-tested. But such prospects are dependent on the availability of sufficient volumes of clean liquid and gaseous biofuels needed to power these vehicles during the technology transition period.

13.4.2.2.3 Industrial processes and products

The manufacturing industry generates emissions in two ways: the GHG produced by the energy used in their production (amounting to 24% of the total global emissions), and direct production of GHG by the manufacturing process itself (accounting for 5.2% of the global total) [22].

Two direct producers of emissions are the cement industry (3.0%) and the chemicals/petrochemicals industry (2.2%). The largest emissions arising from industrial energy use during the production of commodities come from the iron and steel industry (7.2%) and chemicals and petrochemicals (3.6%).

The challenge is to find ways and means of effectively reducing emissions from these commodities.

13.4.2.3 The opportunities

The opportunities for overcoming the above challenges include policy initiatives ranging from high level treaties and agreements to national and international policies. They also include advancements in energy technologies both on the hardware and fuel fronts. The following provides illustrative examples from the power generation, transportation and fuel sub-sectors of the energy industry.

13.4.2.3.1 Coal-fired power generation

Two methods for reducing emissions due to coal-fired power generation are the phasing out of coal-fired power generation and reducing emissions from coal use.

Support for the first was received at COP26, when a Commitment (through a *Global Coal to Clean Power Transition Statement*) was made by 190 nations, regions and organisations to phase-out coal-fired power stations [27].

Coal power stations are sources of despatchable energy, and are thus essential components of electrical grid systems. Removing them from the grid would necessitate the introduction of components that can play an equivalent role. This can be provided, for instance, by energy storage, of which pumped hydro and battery storage are of primary interest.

Currently, the latter consists of large capacity lithium-ion rechargeable batteries which are expensive, providing a strong deterrent to progress. One must await sodium-ion rechargeable batteries to become fully market ready. In addition, **Regenerative Hydrogen Fuel Cells (RHFCs)** are being developed that may be able to satisfy the same storage functions in grids (see chapter 11, section 11.6). However, their roundtrip efficiency needs to increase significantly over the current 30% for them to become viable candidates.

Use of CCUS to capture and store carbon:

A second option for mitigating emissions due to coal-fired power stations is to capture and store (and perhaps use) the emitted carbon.

An example of a proposed carbon capture and storage (CCS) project is the CarbonNet project by Earth Resources that will capture the carbon dioxide released from industries in the La Trobe Valley for storage offshore in nearby Bass strait. This project will be operational by 2030 [28]. The **Australian Government Clean Energy Regulator** has set up an Emissions Reduction Fund that enables emitting organisations to obtain **carbon credits** for reducing their carbon emissions [29].

Unfortunately, it appears that in the case of Australia, the projects that have been tried so far have had limited success. An example is that of the Western Australian LNG gas processing company reported by the Guardian [30] which (evidently) failed to meet its target of capturing the carbon dioxide released from its operations, and had to buy carbon credits to make up for the shortfall. All this indicates the need for more research and development in CCS and CCUS.

13.4.2.3.2 Transportation

Opportunities for reduction of emissions due to transportation involve the three methodologies of policy-making, (hardware) technology innovation and the development and use of clean fuels.

The first two alternatives consist of the phasing out fossil fuel-powered ICE vehicles and the acceleration of the transition to ZEVs. The third alternative includes the use of hydrogen fuel cell vehicles, and the continued use of low-emissions biofuel-powered ICE vehicles.

The ZEV Transition Council (ZEVTC) was formed in November 2020 to accelerate the global transition towards ZEVs. At the COP26 meeting, it announced its Action Plan for 2022 [31]. This included charging infrastructures, and accelerating the pace of the transition to zero carbon for heavy duty vehicles and ensuring the ZEV transition is truly global by bringing together a taskforce to coordinate and deliver development assistance.

The option on the use of clean biofuels in existing ICE vehicles to bridge the possible technology gap due to slow transition towards ZEVs remains largely unexplored.

13.4.2.3.3 Hydrogen

Hydrogen was introduced as a fuel for Fuel Cell EVs (FCEVs) and grid energy storage in chapter 11. Clean hydrogen may be produced through the **electrolysis of water**, if the electricity is generated entirely from renewable energy sources. The resulting hydrogen has relatively little or no emissions associated with both its production and use, and is designated as green hydrogen.

The bulk of hydrogen produced today is through the **steam–methane reforming** or **partial oxidation** of natural gas [32] as described in the chapter 11. These processes produce direct emission of carbon dioxide, and **Carbon dioxide Capture and Storage** (CCS) techniques are possible ways of removing some or most of this emission. The cleaner version of hydrogen is sometimes distinguished from the original product as blue and grey hydrogen, respectively (see chapter 11 for more extended discussions on the nomenclature used to describe the two forms of hydrogen).

Hydrogen can also be used as a domestic fuel source for cooking and heating. It may substitute for natural gas in the form of a blend or in the pure state. In the USA, hydrogen is stored and delivered for such purposes to the end user by either road transportation or pipelines [33].

Gaseous hydrogen requires compression before delivery. This is achieved using reciprocating, rotary, ionic or centrifugal compressors, followed by transportation by gas pipelines or road transportation using tube trailers. *Storage and delivery of hydrogen in liquid form* requires the liquefaction of the gas, which is achieved by reducing the temperature to −253 °C. This is followed by transportation via super-insulated, cryogenic tanker trucks.

In the USA, 1600 miles of dedicated pipelines exist for hydrogen delivery in the gaseous form. These are owned by merchant hydrogen producers and located around the sites of large hydrogen users only. An alternative means of delivery is via hydrogen/natural gas blends using existing natural gas pipelines [34].

Current research in the USA concentrates on issues relating to the use of such pipelines for hydrogen transport. The issues investigated include:

- possible embrittlement of the steel and welds in the pipeline;
- hydrogen permeation and leaks;
- appropriate hydrogen compression technology.

The NREL has carried out a review of the research addressing these issues [35].

The key points of their report are as follows:

(a) If blended at relatively low concentrations (5–15% v/v) with natural gas, hydrogen delivery to end users appears to be viable, and does not significantly increase risks related to the use of the gas blend in end-use devices such as household appliances. Neither does it increase risks to overall public safety, or the durability and integrity of the existing natural gas pipeline. However, blend concentrations may vary significantly over different pipelines and natural gas compositions and therefore each case must be assessed on its own merit.

(b) Blending can also be used to deliver pure hydrogen to markets, with separation and extraction downstream close to the point of use. This avoids the costs of building dedicated hydrogen pipelines. However, the blending and extraction incurs additional costs, as does any modifications to existing pipeline integrity management systems.

(c) Maximum hydrogen blend levels at which no or minor modifications would be needed for end-use systems, including appliances such as household boilers or stoves, fall within the 5–20% hydrogen range.

(d) At the normal pressures and stress levels in the natural gas distribution system, hydrogen induced failures are not a major concern for steel pipes or other metallic pipes. There is no major concern for a hydrogen aging effect on polyethylene (PV) or polyvinyl chloride (PVC) pipes.

(e) Leakage: the permeation coefficient is higher for hydrogen than for methane in many polymer materials and elastometric seals used in natural gas distribution systems. As pipes have much larger surface areas than the seals, the pipe walls will account for the majority of gas losses. Permeation rates for hydrogen are 4–5 times higher than for methane in typical polymer pipes in the USA. Volume leakage rates are about 3 times higher for hydrogen than for methane.

13.4.2.3.4 Fuels for power generation

Biogas for power generation:

Biogas and biomethane are relatively minor players in the global power generation and transportation industries. In chapter 6, it was noted that biogas produced 0.72% share of the global renewable energy power, and provided 0.08% of the global final energy consumption. However, the potential for future use of these clean fuels is very promising.

As noted in chapter 10, in the year 2018, 31.4% of biogas produced globally was used for power generation, 32.7% for co-generation (CHP), 27.3% for heating and cooking while 8.6% was upgraded to biomethane. While the latter fuel is only a tiny proportion of the total biogas use, it is worthy to note that because of its similarity to natural gas (which is predominantly methane in composition), it can be blended with natural gas in any proportion without any modifications. This means that methane/natural gas blends can be used as a substitute for all natural gas applications. As a consequence, biomethane (which is produced largely from the upgrading of biogas) has been pumped into the natural gas pipelines of many EU countries for at least a decade. The member countries include the Netherlands, Germany, Austria, Sweden and France [36].

Pyrolysis oil:

Pyrolysis oil (or bio oil) is one of the emerging fuels for power generation and transportation that was considered in chapter 11. It has properties that can be very similar to those of HFO (see table 11.1) and is thus a suitable substitute for this fuel used in power plants. In 2020, there were at least six companies that were producing pyrolysis oil commercially by the fast pyrolysis technique in Sweden, Finland, Netherlands and Canada (see table 6.2), and one firm in Finland (Savon Voima) using it as fuel for its CHP plant (see box 6.2).

Pyrolysis oils produced from suitable feedstocks could have fuel properties very similar to diesel, and could thus be used as an alternative fuel for transportation. This is the case with pyrolysis oil produced from tyres and plastics (see research question 11.1).

Ammonia as a fuel:

Ammonia (gross calorific value 22.5 MJ kg^{-1}) has a high hydrogen density, and can be used as fuel for combustion systems, as well as in solid oxide fuel cells. Its first use as a fuel was as an ammonia/coal gas blend in reciprocating engines during World War II, and has recently been successfully co-fired with pulverised coal in gas turbines and industrial furnaces [37].

Some of the challenges of using ammonia as a fuel are [38]:

- high ignition temperature;
- low flame velocity;
- slow chemical kinetics.

Ammonia also provides a suitable energy carrier, and is currently the subject of active research as an alternative fuel.

13.4.2.3.5 Industrial production and trade

Overall, the direct and indirect contributions from the manufacture of commodities amounts to 29.4% of the total global emissions per annum. Two such commodities are iron and aluminium. In 2021, emissions from (the energy used in the production of) iron and steel was 7.2%, while total emissions from aluminium manufacture comprised 3% of the world's direct carbon dioxide emission of 9.1 Gigatonnes [39].

An important tool that countries use to limit such industrial emissions is through **carbon taxes** on carbon-intensive products. Amongst the most carbon-intensive commodities in the manufacturing sector are **cement**, **fertilisers**, **iron and steel**, **aluminium** and **electricity**.

As such products form part of international trade, there is a need to establish international trade agreements and policies to monitor and limit the embedded emissions associated with these products, and to ensure equity between taxes imposed on internal and external producers of such products. An example of such an arrangement is the EU's proposal for a **Carbon Border Adjustment Mechanism (CBAM)** [40].

The next section provides a country-specific insight into how nations are responding to the 'net zero by 2050' call from the UNFCCC, by showing how

Australia is going through its energy transition in order to achieve its Paris Agreement goal.

13.5 The energy transition—an Australian case study

Australia's strategy to achieve net zero emissions by 2050 must be viewed in the light of two considerations—the country's heavy dependence on coal and other carbon-intensive commodities for power generation and export, as well as a change in its government in May 2022, soon after the COP26 meeting in Glasgow.

The former federal government had showed a noticeable reluctance towards the prospects of phasing out coal-fired power generation in the country. This evidently had a significant impact on the approach the government took to putting together the net zero plan that was presented to COP26 in November 2021.

Notwithstanding this limitation, Australia did indeed utilise many of the above-mentioned elements and tools of the global energy transition in the formulation of its own plan. A key feature of the Australian energy transition is the role universities are playing in the required research and development.

This section begins with an overview of Australia's net zero plan, and goes on to exemplify the role of Australian universities in the energy transition with the Australian National University's *Grand Challenge* initiative.

13.5.1 Australia's net zero plan

Australia's long-term emissions reduction plan [41], as prepared by the government that existed prior to May 2022, had a target of 26–28% reductions below 2005 levels by 2030. The approach to be used was essentially to drive down the costs of new and existing low-emissions technologies and to accelerate their deployment at scale over a decade. The country's dependence on carbon-intensive exports was to be addressed through a decarbonisation programme that included the use of renewable energy and carbon capture and storage technologies, as well as the use of carbon offsets.

The priority low-emissions technologies chosen were **clean hydrogen**, **ultra low-cost solar**, **energy storage**, **low-emissions steel and aluminium**, **carbon capture and storage** and **soil carbon**. These were prioritised through a **Technology Investment Roadmap**, to contribute emissions reduction to the overall net zero plan according to the scheme shown in table 13.6.

The scheme assigned 100% emissions to the baseline year of 2005, and acknowledged that 20% reductions had already taken place by the year 2020. It then assigned a further 40% reductions to the use of low-emissions technologies that comprised the Technology Investment Roadmap, 15% to global technology trends such as a shift to EVs, 10% to the use of carbon offsets purchased either domestically or internationally, and the remaining 15% to new and emerging technologies. Examples of the latter were the reduction in methane emissions from livestock through algal supplementation of feedstock and the production of low-emissions cement with carbon capture, use and storage.

Table 13.6. Contributions of priority low-emissions technologies towards Australia's net zero plan (data source: [41]).

Australia's plan (before 2022) to achieve its targeted emissions reductions by 2030					
2005 Baseline emission	Reduction up to 2020	Technology Investment Roadmap	Global technology trends	International and domestic carbon offsets	Further technology breakthroughs
100%	20%	40%	15%	10%	15%

To implement the Roadmap, the government was to invest $20bn over the next decade to unlock a further $80bn of private and public investments in these low-emissions technologies.

As evident from earlier discussions, coal-fired power plants contribute an enormous amount to the total carbon emissions globally. In Australia's case, 54% of national grid electricity comes from such power plants. Prior to May 2022, the government had been extremely reluctant to phase-out such power plants. With the arrival of the new government after May 2022, a distinct policy shift took place which was more conducive towards the phasing out of coal power and better disposed towards the ultimate goal of the Paris Agreement. A consequence was the passage in parliament of the **Climate Change Bill of 2022**. Amongst other things, it increased Australia's pledge of emissions reductions by 2030 to 43% above the 2005 levels. The manner in which this increase is to be implemented is yet to be announced.

13.5.2 The role of Australian universities—ANU's grand challenge

As suggested above, Australian universities have been instrumental in contributing substantially to the country's net zero plan. An example is the Australian National University (ANU), which launched its own programme, popularly known as the **Grand Challenge**, in 2019, towards the effort. Box 13.3 provides a more detailed description of this programme of research and engagement.

Box 13.3. ANU's Grand Challenge programme.

This $10 million strategic research programme, more formally known as the **Zero-Carbon Energy for the Asia-Pacific (ZCEAP)** programme [42] is based on a rationale that utilises the abundance in Australia of both renewable energy and mineral resources, and the availability of clean energy to extract and refine the resources.

Australia has a large export trade with the Asia-Pacific region, with more than 90% of its exports of iron ore, coal, petroleum gas, gold and frozen bovine meat (valued at

$162.3 bn) going to China, Japan, South Korea and India [43]. The country also generates 54% of its electricity from coal. With its vast renewable energy supply, Australia has the capacity to decarbonise and export low-emissions energy and commodities to the region.

At a time when the rest of the world is already decarbonising, there is a compelling need for Australia to shift from carbon-emitting exports to zero-carbon exports, and it has the clean energy resources to do so.

The goal of the programme is:

- to support a timely, just and sustainable transition to zero-carbon energy in the Asia-Pacific.

Its objectives are to:

- transform the way Australia trades with the world through the development of zero-carbon export industries;
- create new paradigms in benefit-sharing [44];
- develop technologies, policies and approaches that can be applied in the Asia-Pacific and beyond.

The actual research work initially consisted of the five inter-related projects:

- renewable energy systems;
- hydrogen fuels;
- renewable energy policy and governance in the Asia-Pacific countries;
- renewable metal refining;
- indigenous engagement with renewable energy industries.

A sixth project, titled 'Regulatory frameworks for renewables-based trade and investment program', was added later.

Table 13.7 lists the objectives of the projects with examples typifying the outputs obtained so far for the first five of these projects.

Table 13.7. The projects comprising the ANU Grand Challenge and typical outputs (data source: [42]).

	Grand Challenge Project	Outputs
1	***Renewable Energy Systems:*** • To understand the geopolitical, economic and technological drivers that ensure successful trade in renewable energy • To understand the scale and opportunity for renewable energy electricity exports from North West Australia	1. Cheng C *et al*, 100% Renewable energy in Japan (*Energy Convers. Manag.*) [45] 2. Silalahi *et al*, Indonesia's pumped hydro potential (*Energies*) [46]
2	***Hydrogen Fuels:*** • To enable Australia to become a major exporter of zero-carbon renewable fuels based on hydrogen within the next 15 years	3. Longdon *et al*, Clean hydrogen? (*Appl. Energy*) [47] 4. White *et al*, Emissions certification (*Energy*) [48]

	• To support the application of hydrogen energy use in Australia including the transport sector	5. O'Neill *et al*, RE on 1st nation land (*Aust. Env. Rev.*) [50] 6. Gaykema *et al*, ACT natural gas network (*Int. J. Hydrog. Energy*) [51]
3	***Renewable Energy Policy and Governance in the Asia-Pacific countries:*** • To establish the policy and governance frameworks for zero-carbon energy in the Asia-Pacific region	7. Burke *et al*, Regional decarbonisation (*Energy*) [44] 8. Burke and Do. Greening Asia's Development (*Asian Econ. Policy Rev.*) [52]
4	***Renewable Energy Metal Refining:*** • To develop the technological, policy and market pathways for a new Australian industry in renewably-refined metals, especially iron	9. Venkataraman *et al* (*Energy Policy*, 2022) [49]
5	***Indigeneous engagement with renewable energy industries:*** • To develop the policy and financial scaffolding that ensures multi-generational benefits to indigenous people from North West Australia from the development of renewable energy and associated industries on their land.	10. Riley, Scaling up (Nulungu Publication) [53] 11. O'Neill *et al*, RE on 1st nation land (*Aust. Env. Rev.*) [50] 12. O'Neill *et al* (*Energy Res. Soc. Sci.*, 2021) [54]

In summary, this research initiative takes a holistic and multi-disciplinary approach to decarbonising Australia's exports to the Asia-Pacific region. The chosen approach involves policy development on the regional scale and critical technology developments that are essential for the achievement of the final goal. A range of technologies are considered, with hydrogen playing a key role. The socio-cultural impact of clean energy technology developments on indigenous people's land are elaborated on and the need for equitable benefit-sharing is stressed.

13.5.3 Summary remarks

The Australian government's net zero plan shows how a country's response can be politically driven, with a final outcome that is determined by stakeholders who may not be primarily motivated by concern for the long-term impacts of climate change. Perhaps as a result of this, there was no mention of the phasing out of coal-fired power generation in the country in the original plan as prepared by the previous government. To its credit however, the plan does introduce:

- clean hydrogen as a new fuel and energy storage medium (for intermittent renewable energy and fuel cells);
- energy storage devices (in particular battery storage for grid and stand-alone power);
- low-emissions steel and aluminium;
- carbon capture and storage (including storage in mine sites and through afforestation);
- carbon markets (for carbon offsets acquired both domestically and from abroad);

all of which are in keeping with global trends in energy transition.

Changes (in both the approach and details) to the net zero plan are anticipated with the arrival of the new government.

In contrast, the Australian National University's programme is much better aligned with the new tools and opportunities identified in section 13.4.2.3. This is revealed in a comparison of the outputs shown in table 13.7 with the list of new tools and opportunities discussed in section 13.4.2.3 above.

To achieve its aim, a decarbonisation programme such as ANU's must incorporate substantial engagement with the stakeholders as a mandatory requirement. It must also contain the three key elements of policy, standards and relevant technology.

Towards these ends, the paper by Burke on regional decarbonisation [44] provides an overview of the nature and scale of the challenge, and underpins the policy requirements. New technologies such as hydrogen have been investigated from multiple and multi-disciplinary perspectives. Critical to the success of any decarbonisation scheme involving renewable energy are the socio-cultural factors of land ownership and traditional rights. These dimensions of the energy transition have also received some attention in the ANU outcomes.

Of special note is a paper by Rafiee *et al* on biogas [55]. This highly relevant work reviews the potential of biogas (and by inference biomethane obtained from its purification) as another potential source of clean energy.

An important opportunity that has not been discussed in the ANU research is the potential role of pyrolysis oil as future fuel for power generation and transportation.

(See question 13.2—The role of biogas/biomethane in future energy technology.)
(See research question 13.2—The uncertain dynamics of the energy transition.)

13.6 COP27

In section 13.4, we saw how COP26 set the agenda for a global energy transition to limit the global temperature increase to 1.5 °C above pre-industrial levels. One of the explicit strategies was to phase-out coal-fired power generation. What happened at COP27, and how successful was it in meeting the targets set in the earlier COP?

COP27, which took place at *Sharm el-Sheikh* in Egypt during the period 6–18 November (later extended to 20 November), was known as the 'implementation COP', as its main function was to implement the goals of COP26 and other pending goals of the UNFCCC. The COP27 Agenda [56] contained 20 items, of which the new theme of **Loss and Damage** that appeared under agenda item 7 was of special interest.

This issue is the subject of Article 8 of the Paris Agreement [2], and refers to the 'loss and damage associated with adverse effects of climate change, including extreme weather events and slow onset events'. Essentially, it calls for the wealthy nations (i.e. the developed country Parties to the conference) to compensate the poor countries (the developing country Parties) for losses incurred to them by climate change impacts. As noted below, this agenda item did have a degree of success. But this was not the case with the more pressing issues relating to emissions reductions, including the phase-out of fossil fuels.

A summary statement of the outcomes of COP27 is contained in the UN Climate Press Release of 20 November 2022 [57]. Unfortunately, this statement fails to clearly and adequately stress the main concern regarding the outcomes. Titled 'COP27 reaches breakthrough agreement on new 'Loss and Damage' fund for vulnerable countries', this release correctly portrays the outcomes of the Loss and Damage negotiations. However, it plays down the disappointing results relating to the other key issues discussed, including limiting the global temperature rise to 1.5 °C.

It does, however, include the reminder (in paragraph 18 of the release) by the UN Climate Change Executive Secretary, Simon Stiell, of the IPCC's findings that GHG emissions had to decline by 45% by 2030 to limit global warming to 1.5 °C. (This was revealed in the UN Climate Press Release, titled 'Climate plans remain insufficient: More ambitious actions needed now' of 26 October 2022 [58].)

A more transparent and factual reporting of the key outcomes of COP27 is provided by ***The Conversation*** newsletter [59].

According to this newsletter, the urgent actions required at the meeting were:

- support for people displaced by extreme events due to climate change (i.e. Loss and Damage compensation);
- transparent governance of carbon markets;
- the phase-out of **all** fossil fuels.

The significant achievement was the establishment of a new fund for loss and damage finance. This was to be paid by the developed country Parties to the developing country parties (which consisted of the G77 countries and China). The Parties agreed to set up a **transitional committee** to make recommendations for adoption at COP28 in November 2023. The new climate finance goal, that replaced the $US 100 billion annual target, was delayed to next year (2023).

According to this report [59], on the issue of phasing out fossil fuels:

- COP27 **failed** to obtain a commitment from all parties to phase-out all fossil fuels.
- The meeting only agreed to the wording 'accelerating efforts towards the phase-down of unabated coal power and phase-out of inefficient fossil fuel subsidies' in the description of the outcomes.

The term 'unabated' evidently referred to coal-fired power stations without CCS. This strongly implies the prospects of the continued use of coal-fired power stations that 'abate' their emissions via the use of CCS.

In summary, COP27 failed dismally to respond to the call for urgent climate action that the scientific reports had been indicating as necessary since well before the previous COP meeting.

Exercises and questions

Exercises

Exercise 13.1. Tracing the history of the global warming targets.

(a)
  (i) When (and where) did the Parties of the UNFCCC **first** agree to limit global warming to 2 °C?
  (ii) When did 1.5 °C become more significant?

(b) The Paris Agreement aimed to achieve a balance between anthropogenic emissions by sources and removal by sinks of greenhouse gases **in the second half of this century**. When did this deadline change to 2050, and how did this happen?

Exercise 13.2. Comparing Wein's Law and Stefan–Boltzmann's law.

Wein's Law says that the peak wavelength of radiation is inversely proportion to the absolute temperature, while the Stephan–Boltzmann Law relates the power (energy per unit time) to the temperature.

Show that the Stefan–Boltzmann Law is consistent with Wein's law. That is, it shows that the peak wavelength shifts to smaller values as the temperature rises. (Hint: Use the fact that the energy of radiation is inversely proportional to its wavelength.)

Exercise 13.3. Why a GHG increases the Earth's temperature and causes global warming.

Introducing a GHG into the atmosphere reduces the OLR by essentially reflecting back towards the Earth some of the outgoing long wavelength radiation from it. The ASR remains the same.

Use the Earth's energy budget statement and Stefan–Boltzmann's Law to show why there must be a rise in the Earth's temperature (i.e. global warming) when a GHG is introduced to the atmosphere.

Questions and research questions

Question 13.1. Understanding the goal of the Paris Agreement.

The Paris Agreement refers to 'achieving a balance between emissions by sources and removals by sinks', and 'peaking of greenhouse gas emissions'.

(a) What does 'peaking of greenhouse gas emissions' mean in Article 4? Does it apply to a single Party, or to the global scenario?
(b) How is the phrase 'balance between emissions by sources and removals by sinks' commonly rephrased today?
(c) Identify some sources and sinks of greenhouse gases.
(d) What are the three greenhouse gases (GHGs) of most concern today?
(e) What will happen to the level of GHGs in the Earth's atmosphere when a global balance between emissions and removals is finally achieved?
(f) Will it matter if this balance is achieved later rather than sooner?

Question 13.2. The role of biogas/biomethane in future energy technology.

Biogas and biomethane are mature fuel technologies for the power generation and transportation sectors. In chapter 6, it was noted that the current biogas production is equal to only 1.6–2.2% of total global potential for this biofuel.

(a) Review the sources of biogas feedstocks presented in the World Biogas Association report (Global potential for biogas, June 2019), and comment on how this could be used to increase biogas production globally.
(b) Discuss the conditions under which biogas/biomethane will become significant contributors to the ongoing energy transition (i.e. what must be true before these fuels become important as fuels for transportation and power generation in the future?).

Research question 13.1. Motivation for the structure and contents of the COP26 goals.

The goals of COP26 are presented under four separate headings, at least some of which are evidently related to reports and events that preceded this 2021 meeting in Glasgow.

Carry out an analysis of these and similar climate-related activities that could have influenced, or determined the formulation of the goals of COP26. Your answer must address at least two of the thematic questions below.

(a) How many of the goals were carried over from previous COPs? Expand your answer with details.
(b) Are any of the goals related to the IPCC Special Report on 1.5 °C, the Energy Gap Report or any other issue or event that occurred prior to the Glasgow meeting?
(c) How successful was COP26 in achieving the desired outcomes of its goals?
(d) In your assessment, were there any other goals that could have been presented for discussions at the Glasgow meeting?

Research question 13.2. The uncertain dynamics of the energy transition.

The arrival of zero emission vehicles (ZEVs) as mainstream transportation will seriously diminish the importance of internal combustion engine (ICE) vehicles.

However, it is not certain if, and when, ZEVs will take over road transportation completely. Some countries have already announced policies and legislation to ban certain categories of ICE vehicles in favour of ZEVs in the future.

(a) Investigate the global status of policies relating to ZEVs. Which countries have placed a future ban on the use of ICE vehicles in preference to ZEVs, and when do these legislations become effective?
(b) Ascertain the current status of liquid and gaseous biofuels as transportation fuels.
   - Which types of biofuels are being used now, and how much of each is being produced?
   - What is the present share of these fuels on the transportation fuel market?
(c) Think of the time interval between now and the arrival of the ZEVs. The countries of the world can still reduce their emissions using ICE vehicles during the interim period. For this, they will need to transition from fossil fuel powered ICE vehicles to biofuel-powered vehicles.

   How much biofuel will be required for such a transition? Consider the current global biofuel supply, and estimate the time that will be needed for it to reach the required level.

Are there other ways in which a zero-carbon transport industry could be achieved? Suggest at least one such possible scenario.

References

[1] From Stockholm to Kyoto: a brief history of climate change. UN [accessed 11 September 2021]. Available from https://un.org/en/chronicle/article/stockholm-kyoto-brief-history-climate-change
[2] Paris Agreement. UNFCCC [accessed 14 September 2022]. Available from https://unfccc.int/sites/default/files/english_paris_agreement.pdf
[3] Nationally Determined Contributions Registry. UN Climate Change [accessed 10 September 2022]. Available from https://unfccc.int/NDCREG
[4] Special Report: Global Warming of 1.5 °C. IPCC [accessed 10 September 2022]. Available from https://ipcc.ch/sr15/about/
[5] SDG Knowledge Hub. Decoding net-zero pledges to meet the Paris Agreement Goals, SDG7 [accessed 10 September 2022]. Available from https://sdg.iisd.org/commentary/policy-briefs/decoding-net-zero-pledges-to-meet-paris-agreement-goals-sdg-7/
[6] UN High Level Dialogue on Energy. 24 September 2021 [accessed 10 September 2022]. Available from https://un.org/en/conferences/energy2021
[7] Emissions Gap Report 2021. UNEP, UNEP Copenhagen Climate Center [accessed 26 October 2021]. Available from https://unep.org/resources/emissions-gap-report-2021
[8] Achievements of the Conference of the Parties. Sustainability for All [accessed 11 September 2021]. Available from https://activesustainability.com/climate-change/achievements-of-the-conference-of-the-parties/?_adin=02021864894

[9] NASA Earth Observatory. Climate and Earth's Energy Budget [accessed 14 September 2022]. Available from https://Earthobservatory.nasa.gov/features/EnergyBalance
[10] Earth's Energy Budget. Wiki [accessed 14 September 2022]. Available from https://en.wikipedia.org/wiki/Earth%27s_energy_budget
[11] Stefan Boltzmann Law. Britannica [accessed 14 September 2022]. Available from https://britannica.com/science/Stefan-Boltzmann-law
[12] NOAA Annual Greenhouse Gas Index. NOAA Global Monitoring Laboratory [accessed 14 September 2022] Available from https://gml.noaa.gov/aggi/aggi.html
[13] Climate Sensitivity. ACS Climate Science Toolkit. ACS Chemistry for Life [accessed 14 September 2022]. Available from https://acs.org/content/acs/en/climatescience/atmospheric-warming/climatsensitivity.html or other authoritative sources
[14] Greenhouse gas protocol. Global warming values [accessed 14 September 2022]. Available from https://ghgprotocol.org/sites/default/files/ghgp/Global-Warming-Potential-Values%20%28Feb%2016%202016%29-1.pdf
[15] 2006 IPCC Guidelines for National Greenhouse Gas Inventories Vol 1. IPCC [accessed 15 September 2022]. Available from https://ipcc-nggip.iges.or.jp/public/2006gl/vol1.html
[16] 2019 Refinement to the 2006 IPCC Guideline. IPCC [accessed 15 September 2022]. Available from https://ipcc.ch/report/2019-refinement-to-the-2006-ipcc-guidelines-for-national-greenhouse-gas-inventories/
[17] Mani F S 2020 Estimating greenhouse gas emission in the Pacific Island Countries ed A Singh and R C Deo *Translating the Paris Agreement into Action in the Pacific* (Cham: Springer Nature), pp 3–27
[18] COP26 Goals. UN Climate Change Conf. UK 20201 [accessed 18 October 2022]. Available from https://ukcop26.org/cop26-goals/
[19] Annual greenhouse gas emissions. Our World in Data [accessed 18 September 2022]. Available from https://ourworldindata.org/greenhouse-gas-emissions for selected years between 2000 and 2021
[20] Updated synthesis report: worrying trends confirmed. UNFCCC [accessed 18 September 2022] Available from https://unfccc.int/news/updated-ndc-synthesis-report-worrying-trends-confirmed
[21] Emissions by Sector. Our World in Data [accessed 14 September 2022]. Available from https://ourworldindata.org/emissions-by-sector
[22] Top 25 coal power countries in 2020. EMBER [accessed 20 September 2022]. Available from https://ember-climate.org/insights/research/top-25-coal-power-countries-in-2020/
[23] Emissions. IEA [accessed 20 September 2022]. Available from https://iea.org/reports/global-energy-co2-status-report-2019/emissions
[24] Transport 2020 IEA [accessed 20 September 2022]. Available from https://iea.org/reports/tracking-transport-2020
[25] Abraham K M 2020 How comparable are sodium-ion batteries to lithium-ion counterparts *ACS Energy Lett.* **5** 3544-47 Available from https://pubs.acs.org/doi/10.1021/acsenergylett.0c02181
[26] The weekend read: sodium-ion batteries go mainstream. *PV Magazine* [accessed 22 September 2022]. Available from https://pv-magazine.com/2022/03/26/the-weekend-read-sodium-ion-batteries-go-mainstream/
[27] COP26: 190 nations, regions and organisations commit to major coal power and financing phase-outs. Euractiv [accessed 22 September 2022]. Available from https://euractiv.com/

section/energy/news/cop26–190-nations-regions-and-organisations-commit-to-major-coal-power-and-financing-phase-outs/

[28] The CarbonNet Project. Earth Resources [accessed 26 September 2022]. Available from https://Earthresources.vic.gov.au/projects/carbonnet-project/about-the-project

[29] Carbon capture and storage method. Australian Government Clean Energy Regulator [accessed 26 September 2022]. Available from https://cleanenergyregulator.gov.au/ERF/Choosing-a-project-type/Opportunities-for-industry/carbon-capture-and-storage-method

[30] Australia's only working carbon capture and storage project fails to meet target. *The Guardian* [accessed 18 October 2022]. Available from https://theguardian.com/australia-news/2021/nov/12/australias-only-working-carbon-capture-and-storage-project-fails-to-meet-target

[31] Zero Emission Vehicles Transition Council: 2022 action plan COP26 [accessed 20 September 2022]. Available from https://ukcop26.org/zero-emission-vehicles-transition-council-2022-action-plan/

[32] Hydrogen production: Natural Gas Reforming. Energy.gov [accessed 26 September 2022]. Available from https://energy.gov/eere/fuelcells/hydrogen-production-natural-gas-reforming

[33] Hydrogen delivery. Energy.gov [accessed 25 September 2022]. Available from https://energy.gov/eere/fuelcells/hydrogen-delivery

[34] Hydrogen pipelines. Energy.gov9 [accessed 26 September 2022]. Available from https://energy.gov/eere/fuelcells/hydrogen-pipelines

[35] Melaina M W, Antonia O and Penev M Blending hydrogen into natural gas pipeline networks: a review of key issues. NREL. 2013 [accessed 26 September 2022]. Available from https://nrel.gov/docs/fy13osti/51995.pdf

[36] Savickisl J, Zemite L, Zeltins N, Bode I, Jansons L, Dzelzitis E, Koposovs A, Selickis A and Ansone A 2020 The biomethane injection into the natural gas networks: the EU's gas synergy path *Sciendo* **4** 34–50 Available from https://researchgate.net/publication/343716958_The_Biomethane_Injection_into_the_Natural_Gas_Networks_The_EU%27s_Gas_Synergy_Path

[37] Kobayashi H, Hayakawa A, Kunkuma K D, Somarathne A, Ekenechukwu and Okafor C 2019 Science and technology of ammonia combustion *Proc. Combust. Inst.* **37** 109–33 Available from https://sciencedirect.com/science/article/pii/S1540748918306345

[38] Erdemir D and Dincer I 2021 A perspective on the use of ammonia as a clean fuel:challenges and solutions *Int. J.Energy Res.* **45** 4827–34

[39] Aluminium. IEA [accessed 25 September 2022]. Available from https://iea.org/reports/aluminium

[40] Carbon border adjustment mechanism: questions and answers. *European Commission* [accessed 22 September 2022]. Available from https://ec.europa.eu/commission/presscorner/detail/en/qanda-21-3661

[41] Australia's long-term emissions reduction plan. Australian Government Department of Climate Change, Energy, the Environment and Water [accessed 19 October 2022]. Available from https://dcceew.gov.au/climate-change/publications/australias-long-term-emissions-reduction-plan

[42] Zero Carbon energy for the Asia-Pacific (ZCEAP). Australian National University [accessed 29 September 2022]. Available from https://anu.edu.au/research/research-initiatives/zero-carbon-energy-for-the-asia-pacific

[43] OEC Profiles. OEC [accessed 28 September 2022]. Available from https://oec.world/en/profile/country/aus

[44] Burke P J, Beck F J, Aisbett E, Baldwin K G H, Stocks M, Pye J, Venkataraman M, Hunt J and Bai X 2022 Contributing to regional decarbonization: Australia's potential to supply zero-carbon commodities to the Asia-Pacific *Energy* **248** 123563

[45] Cheng C, Blakers A, Stocks M and Lu B 2022 100% Renewable energy in Japan *Energy Convers. Manag.* **255** 115299

[46] Silalahi D F, Blakers A, Lu B and Cheng C 2022 Indonesia's vast off-river pumped hydro energy storage potential *Energies* **15** 3457

[47] Longden T, Beck F, Jotzo F, Andrews R and Prasad M 2022 Clean hydrogen?—comparing the emissions and costs of fossil fuel versus renewable-based hydrogen *Appl. Energy* **306** 118145

[48] White L, Cheng W, Aisbett E, Beck F, Baldwin K G H, Howarth P and O'Neill L 2021 Towards emissions certification systems for international trade in hydrogen *Energy* **215** 119139

[49] Venkataraman M, Csereklyei Z, Aisbett E, Rahbari A, Jotzo F, Lord M and Pye J 2022 Zero-carbon steel production: the opportunities and role for Australia *Energy Policy* **163** 112811

[50] O'Neill L, Beck F, Cheng W and Nolan K 2022 Renewable hydrogen produced on land owned by first nation people- will the owners benefit? *Aust. Environ. Rev.* 149–58

[51] Gaykema E, Skryabin I, Prest J and Hansen B 2021 Assessing the viability of the ACT natural gas distribution system for re-use as hydrogen distribution network *Int. J. Hydrog. Energy* **46** 12280–9

[52] Burke P and Do T 2021 Greening Asia's economic development *Asian Econ. Rev. Policy* **16** 22–39

[53] Riley B 2021 Scaling up: renewable energy on aboriginal lands in North West Australia. Nulungu Publications Series *Nulungu Research Paper* **3** 1–24

[54] O'Neill L, Thorburn K, Riley B, Maynard G, Shirlow E and Hunt J 2021 Renwable energy development on the indigeneous estate: free, prior and informed consent and best practice in agreement-making in Australia *Energy Res. Soc. Sci.* **81** 102252

[55] Rafiee A, Khalipour K, Prest J and Skryabin I 2021 Biogas as an energy vector *Biomass Bioenergy* **144** 105935

[56] COP27. UN Climate Change [accessed 23 November 2022]. Available from https://unfccc.int/event/cop-27

[57] COP27 reaches breakthrough agreement on new 'Loss and Damage' fund for vulnerable countries. United Nations Climate [accessed 23 November 2022]. Available from https://unfccc.int/news/cop27-reaches-breakthrough-agreement-on-new-loss-and-damage-fund-for-vulnerable-countries

[58] Climate plans remain insufficient: more ambitious actions needed now. UN Climate Change [accessed 23 November 2022]. Available from https://unfccc.int/news/climate-plans-remain-insufficient-more-ambitious-action-needed-now

[59] COP27 key outcomes: progress on compensation for developing countries, but more needed on climate justice and equity. *The Conversation* [accessed 23 November 2022]. Available from https://theconversation.com/cop27-key-outcomes-progress-on-compensation-for-developing-countries-but-more-needed-on-climate-justice-and-equity-195017

IOP Publishing

Anirudh Singh

Chapter 14

Energy, pandemics and an integrated global development plan

Energy is needed to facilitate global economics. Improvement in the economic state of the individual improves human well-being. There are several global challenges to human well-being today. They include crises and disasters such as climate change, poverty and its consequences, pandemics, earthquakes, storms and draughts. These are currently being addressed under separate policy frameworks.

The aim of this chapter is to gain better insights into the nature of these challenges, and to suggest a possible improvement to the methodology used to address them. The chapter first surveys the Millennium Development Goals (MDGs) and the Sustainable Development Goals (SDGs). It then examines the Sendai Framework for Disaster Risk Reduction (SFDRR) in more detail to ascertain the methodologies currently being employed for addressing disasters. A possible way of establishing an integrated approach for all global challenges is to adopt a single framework for addressing them all together. A procedure for establishing such an integrated global strategy is suggested.

14.1 Introduction

Energy is a vital enabler of global economics, and therefore has a direct influence on human well-being. There are many global challenges that have an impact on human well-being today. These may be divided into global crises and disasters which are either human-made or natural, and physical or biological.

Climate change is an obvious example of a human-made crisis, causing disasters such as increased occurrences of extreme weather events (storms, floods and draughts) coastal inundation, polar ice melts and sea-level rise. Poverty, the lack of basic economic resources, is another example of a global crisis. Most recently, the world was afflicted by COVID-19, which may be categorised as a biological crisis.

doi:10.1088/978-0-7503-3555-3ch14

Disasters that are exclusively natural and involve no human influence include earthquakes and tsunamis.

It is interesting to note that all these challenges have been addressed in the same way: through policy frameworks and action plans emanating from the highest levels of decision-making, the United Nations or its agencies. Thus, the issues of climate change, global poverty, disasters and COVID-19 are being considered by the UNFCCC, SDGs, the Sendai Framework for Disaster Risk Reduction and the WHO, respectively.

What is most noticeable about these global challenges is that each has been addressed using specific strategies devised according to its own needs. At first sight, a common solution does not seem to be possible, as there is seemingly no common denominator that links these separate challenges to human well-being. Yet it may be possible to find such a common element, and to use it to develop an integrated framework to address all the issues together.

This chapter begins by surveying the millennium development goals (MGDs), which set out in 2000 to address global poverty and its consequences, and the Sustainable Development Goals (SDGs) which succeeded the MDGs and expanded them by including the additional global challenges of energy access and climate change. Section 14.3 takes a closer look at how energy access is related to the health and economic well-being of the poor. The next section considers how we are now addressing global crises and disasters by first enumerating and classifying them. It examines the Sendai Framework for Disaster Risk Reduction (SFDRR) as a case study of a particular strategy for addressing disasters.

It is noted that most of the strategies which are currently in place for addressing global challenges have evolved separately, and there is a need for an integrated approach that considers them all together. Section 14.5 makes a suggestion for such an integrated approach, and discusses possible criteria for the development of such a common strategy.

14.2 Global poverty, the MDGs and the SDGs

14.2.1 Poverty and the MDGs

At the end of the Millennium Summit lasting 6–8 September 2000 at the UN headquarters in New York, the United Nations established the following eight **Millennium Development Goals (MGDs)** [1]:

1. Eradicate extreme poverty and hunger.
2. Achieve universal primary education.
3. Promote gender equality and empower women.
4. Reduce child mortality.
5. Improve maternal health.
6. Combat HIV/AIDS, malaria and other diseases.
7. Ensure environmental sustainability.
8. Develop a global partnership for development.

The establishment of these MDGs by the 190 member states of the organisation was the result of a growing concern amongst the world's leaders regarding, amongst other things, the widening gap between the rich and the poor and the resulting inequities. The leading concern of the MDGs was poverty, which is inevitably linked to socio-economic development. The other goals were largely devoted to achieving such an aim.

It is significant to note that access to electricity (and modern energy in general) was not articulated in the MDGs, and neither was the link between energy and poverty. Indeed, there seems to be no reference to energy at all in the millennium declaration.

The realisation of the existence of a demonstrated link between energy, poverty and the quality of life, however, was clearly revealed in statistics that were being collected by social analysts of the time.

According to the World Energy Outlook 2002 [2], in 2000:

- 1.6 billion people (roughly a quarter of the world's population) were without electricity;
- Four fifths of these lived in the rural areas of developing countries, mainly in South Asia and Sub-Saharan Africa;
- 2.4 billion people depended on traditional biomass for their cooking and heating needs.

Clean energy is a necessity of life, and the correlation between poverty and energy access was clear even in this early data. The poorest can only afford the cheapest form of energy (traditional biomass) for their most basic energy needs. This form of energy does provide the poor with their basis energy needs. However, traditional biomass is not clean energy, and harms the health of the user.

The MDGs had several notable successes over their 15-year span, especially in the areas of poverty reduction, child and maternal health and disease control. Yet at the end of 2015, reliable data sources [3] showed that:

- 1.3 billion people still lacked access to electricity;
- 3 billion people relied on traditional biomass; and
- while extreme poverty had reduced from 1.9 billion in 1990 to less than half that amount in 2015, one in five people in developing countries still earned less than $1.25 a day.

The trends in these global indicators of human well-being over the decades straddling the turn of the century are highlighted in table 14.1.

As the table reveals, while both the global number and ratio of people lacking access to electricity was reducing over the period, the same was not true for the numbers who relied on traditional biomass for cooking and heating.

Energy is intricately linked to the basic needs of people, and thus to their socio-economic development. They need [2]:

- electricity for their lighting and household appliances;
- kerosene and LPG to reduce their dependence on traditional biomass for cooking and heating;

- diesel and heavy fuel oil for space heating;
- diesel, gasoline (petrol) and LPG for transportation.

Table 14.1. Global trends in energy access, dependence on traditional biomass and poverty, 2000–15 (data source: [2, 3]).

	2000	2010	2015
Number of people without access to electricity	1.6 billion (one quarter of global population)	1.4 billion (85% in rural areas)	1.3 billion (one-fifth of world population)
Use of traditional biomass for cooking and heating	2.4 billion people	2.7 billion	3 billion
Global extreme poverty	Less than 1.9 billion		Less than 0.9 billion (1 in 5 people in developing countries earn less than $1.25/day)

The inability to sustain these needs demonstrates an inability to maintain a reasonable standard of living, the hallmark of poverty. The term 'energy poverty' is commonly used to indicate the difficulty that people have in providing for their basic energy needs. There are several ways in which this is defined, but perhaps the most common is the percentage of total earnings spent on basic energy needs (the greater the percentage, the poorer the state of the family or individual).

As box 14.1 reveals, energy poverty is prevalent in many regions of the world, including the EU.

Box 14.1. Energy poverty in Europe.

Energy poverty is not confined to regions such as Africa and South Asia alone, but is an issue of concern to the European sub-continent as well [4].

The causes in the EU, however, are different. They range from low personal income, high house rentals, to poorly designed housing and the climate. Unlike countries in the tropical regions, the Europeans are faced with the persistent need to heat their homes during winter. This requires considerable expenditure that can amount to large fractions of family incomes.

In an EU-wide survey carried out in 2019, it was found that 6.9% of the EU population could not afford to heat their homes adequately. This was caused by both high rentals which depleted earnings and high energy costs. It was exacerbated by the poor insulation and state of repair of buildings.

In an effort to address this multi-dimensional issue, the EU has launched the policy package dubbed 'Clean Energy for All Europeans' [5]. In addition, it has introduced the 'Recommendation on energy poverty (EU) 2020/1563' [6] as part of the renovation wave package.

14.2.2 Energy access and the SDGs

In a desire to continue with the spirit of global development that had inspired the millennium declaration and the MDGs, these goals were replaced by a new programme of action after the MDG target year of 2015 was reached. Amongst other things, this programme specifically noted clean energy as essential for socio-economic well-being, and included climate change impacts as a new goal. Over the years 2015–16, an expanded set of 17 global development goals, called the **Sustainable Development Goals (SDGs)** were articulated and implemented by the UN [7].

These goals

- acknowledged clean energy as an essential requirement for development;
- included the new issue of climate change, the importance of which had already been noted by the UN in the establishment of the UNFCCC during the latter part of the last century.

An overview of the history, development and nature of the UN's sustainable development goals is provided in box 14.2.

Box 14.2. What are the Sustainable Development Goals (SDGs)?

The Sustainable Development Goals (SDGs—also known as **Transforming our world: the 2030 Agenda for Sustainable Development**) are a set of **17 goals** of the United Nations, with **169 associated targets**, that cover the broadest range of sustainable development issues confronting the world today.

A UN site for the SDGs that contains the history of their development is found at [8].

A factsheet summarising the 17 goals and providing their social perspective is found at [3].

The goal that deals with energy is Goal 7.

History and motivation for the SDGs

The idea of SDGs was initiated at the United Nations Conference on Sustainable Development in Rio de Janeiro in 2012. The aim of the Rio conference was to define a set of global goals to address the pressing range of environmental, economic and political problems facing the world.

The SDGs took over from the Millennium Development Goals (MGDs) which were adopted by the UN in 2000 to reduce poverty and other social issues by 2015. The SDGs were agreed to in September 2015.

Note: Two other agreements that were reached in 2015 were

- COP21 in Paris to fight Climate Change;
- The Framework for Disaster Risk Reduction, signed in Sendai Japan in March 2015.

The indicators of sustainable development

Each target of a goal needs to be assessed (i.e. monitored (measured) and evaluated) to ascertain the achievement of the goal. This assessment can be carried out through the use of **indicators** at the **global**, **regional** and **national** levels.

The UN established an **Inter-agency Expert Group (IAEG) on SDG Indicators** (IAEG-SDGs) in March 2015 to propose a global indicator framework, as well as the associated global and universal indicators.

Each SDG therefore has

- a name and short title;
- targets that it wants to achieve;
- indicators that can be measured to assess the achievement of the goal.

Sustainable Development Goal 7

The title of the goal is:

To ensure access to affordable, reliable, sustainable and modern energy for all (short title 'Affordable and Clean Energy').

The targets are:

- by 2030, ensure universal access to affordable, reliable and modern energy services;
- by 2030, increase substantially the share of renewable energy in the global energy mix;
- by 2030, double the global rate of improvement in energy efficiency;
- by 2030, enhance international cooperation to facilitate access to clean energy research and technology, including renewable energy, energy efficiency and advanced and cleaner fossil-fuel technology, and promote investment in energy infrastructure and clean energy technology;
- by 2030, expand infrastructure and upgrade technology for supplying modern and sustainable energy services for all in developing countries, in particular least developed countries, small island developing States, and land-locked developing countries, in accordance with their respective programmes of support.

Goal 7 indicators

The proposed global SDG indicators for Goal 7 are:

- proportion of population with access to electricity;
- proportion of population with primary reliance on clean fuels and technology;
- renewable energy share in the final energy consumption,
- energy intensity measured in terms of primary energy and GDP.

Global database

The IAEG has created a platform called 'the Dissemination Platform of the Global SDG Indicators' which provides access to data on the proposed indicators.

This platform is available at [9].

14.3 Energy and economic well-being

We cannot have economic well-being without energy. More energy, and cleaner energy is required for an improvement in the economic well-being of families. The cheapest form of energy is traditional biomass energy, which is often the only available option for the poor because of their economic status. This section shows how access to clean, modern energy is essential for the improvement of the quality of life of the poor.

14.3.1 Access to clean and modern energy

As economic status improves, the poorest can move away from traditional biomass and can spend more on modern energy to enjoy a modern lifestyle.

This is borne out by the World Energy Outlook (2002) study [2], which shows (in figure 13.1 of the study) how the selection of fuels for various needs changes as one transitions from low-income to higher-income families. The lowest income families use energy for cooking, heating and lighting only. The fuels used are amongst the cheapest (biomass and candles). As the income level increases, energy is used for a greater range of goods and services, including transport, basic appliances, refrigeration and water pumps. The fuels include oil, electricity (mains and battery) and diesel-powered electricity. At the highest income levels, new services such as ICT, air-conditioning and other appliances become available through the use of mains electricity, and bring about greater improvement in the quality of life.

Access to modern energy, in the form of electricity, is a good indicator of economic well-being. As [2] reveals in their figure 13.5, an inverse trend exists between poverty (as measured by the percentage of the population earning below $2/day in a country) and access to electricity among countries of the world. The less poor countries have more access to electricity, and enjoy better economic well-being.

(See question 14.1—Energy and well-being: a personal perspective.)

14.3.2 Link between poverty and traditional biomass energy

Poor people in developing countries rely heavily on traditional biomass, especially for their cooking and heating needs. The World Energy Outlook 2002 [2] shows (in table 13.5 of the study) that in developing countries in the year 2000, there were 2.4 billion people, making up 52% on average of the total population, who depended on traditional biomass for cooking and heating.

The link between poverty levels in developing countries and the use of traditional biomass is clearly revealed in figure 13.11 of [2], which shows how poverty (as measured by the percentage of the population earning below $2/day) is related to the share of traditional biomass in residential use in various countries of the world. As the poverty level increases amongst the countries considered, the share of traditional biomass used in the home increases as well. Countries with the highest (average) family incomes in the study were Poland, Morocco and Algeria, while those with the lowest were Zambia and Nigeria.

14.3.3 Clean energy vs traditional biomass energy

Clean energy is energy that produces low pollution. This includes atmospheric pollution via the release of **particulate matter** [10], as well as gaseous pollution, including the **greenhouse gases**.

Traditional biomass is cheap and the most readily available cooking fuel in developing countries. It therefore provides the default form of energy for the poor in these countries. But it has a range of disadvantages, the most significant being its adverse health effects on women who use it as fuel in **traditional cooking stoves**, and their children who accompany them.

Such traditional stoves are highly inefficient, and emit considerable amounts of particulate matter in the form of smoke and soot as well as toxic gases including **carbon monoxide** and **hydrocarbons**. They are thus generally not classified as a clean energy devices and their use is discouraged, even in poor countries where significant proportions of the populations rely on traditional biomass for cooking because of its affordability and availability.

The use of traditional biomass (in the form of wood) also causes deforestation in surrounding regions of the communities of developing countries. This forest depletion is exacerbated by the employment of the (inefficient) traditional cooking stoves.

(See research question 14.1—Traditional biomass and the health and well-being of women and children in poor countries.)

14.4 Addressing global crises and disasters

14.4.1 What are the global crises and disasters facing humans today?

As humankind was only recently reminded, the list of global crises faced by humans does not end with poverty and climate change. The advent of the coronavirus pandemic in early 2020 was a grim reminder that many other global challenges, some yet to be envisaged, are possible. To start any process of addressing such challenges, it is natural to begin by obtaining a holistic picture of the range of such possible crises.

Some of the well-known global crises and disasters include:

- poverty and its consequences;
- climate change and its impacts;
- disease, including epidemics and pandemics;
- earthquakes, tsunamis and other geophysical disasters.

There are also crises related to the global economy. These include:

- financial crises such as the global financial crisis of 2008, often accompanied by the prospects of global recession;
- energy supply crises such as the oil crisis of 1973 and the recent European energy supply crisis due to the Russian invasion of Ukraine that started on 24 February 2022;
- supply chain crises such as those that are being felt globally since the onset of COVID-19.

Global crises are threats to human well-being that emerge globally (or become global in extent over time), and often manifest themselves in the form of disasters at the local scale. In contrast to global crises, disasters are usually more localised in space and time, often happening suddenly and demanding immediate action. The International Federation of Red Cross and Red Crescent (IFRC) defines a disaster as

'serious disruptions to the functioning of a community that exceed its capacity to cope using its own resources' [11].

Crises and disasters may be divided into human-made or natural. An earthquake is a natural disaster, while a nuclear disaster is appropriately classified as human-made.

(See exercise 14.1—Classifying human crises and disasters.)

14.4.2 What are the strategies used to address global crises and disasters?

Unlike disasters, global crises usually develop slowly, allowing world governing organisations and individual governments time to deliberate on their nature and evolve structured problem-solving methodologies to deal with them. A universally-accepted procedure to address issues of such importance is to:

- establish policy frameworks;
- implement the policies with the assistance of **Action Plans** and **Roadmaps**;
- monitor success using easily-measurable indicators, and review the policies where appropriate.

Usually, it is the world governing organisations that work out the policies and the strategies for implementing them. Implementation is left to individual governments, who are often assisted by NGOs and aid organisations such as the **International Federation of Red Cross and Red Crescent (IFRC)**. In the case of disasters, the latter often play a more important role.

Three major UN Policy frameworks for crises and disaster risk management are:

- The UNFCCC and IPCC for combating climate change;
- The Sustainable Development Goals (SDGs) for poverty and sustainable development;
- The **Sendai Framework for Disaster Risk Reduction (SFDRR)** [12].

Unlike frameworks for global crises, the Sendai Framework for Disaster Risk Reduction is concerned with the impacts on human well-being of disasters, which are short-term events such as earthquakes, draughts, cyclones and flooding, some of which may arise as a result of global crises. Box 14.3 provides a brief introduction to this UN-sponsored framework.

Box 14.3. The Sendai Framework for Disaster Risk Reduction.

The Sendai Framework for Disaster Risk Reduction (SFDRR) was adopted at the **Third UN World Conference on Disaster Risk Reduction**, held at Sendai, Japan during 14–18 March 2015 [13]. It was a follow-up to the **Hyogo Framework for Action**, adopted by the UN in 2005, which had a similar aim of reducing disaster risk (see preamble to [12]).

The Hyogo Framework had made some progress in reducing disaster risk at all of the local, regional, national and global levels, leading to a decrease in mortality due to some hazards. Under this framework, countries had enhanced their capacities for disaster risk management, and international mechanisms such as the **Global Platform for Disaster Risk Reduction** had developed policies and strategies for disaster risk reduction.

However, disasters were continuing to incur a heavy toll on the well-being and safety of persons, communities and countries throughout the world. It was clear that a more 'people-centred preventative approach' to disaster risk reduction was called for [12].

The new framework replacing the Hyogo Framework thus aimed to 'guide multi-hazard management of disaster risk at all levels as well as within and across sectors' [12].

The **goal** of the Sendai Framework that emerged from this predecessor is to

'Prevent new, and reduce existing disaster risk through the implementation of integrated and inclusive economic, structural, legal, social, health, cultural, educational, environmental, technological, political and institutional measures that prevent and reduce hazard exposure and vulnerability to disaster, increase preparedness for response and recovery, and thus strengthen resilience'.

The SFDRR has seven specific **targets**, which may be summarised in essence as follows:

1. Reduce global disaster mortality by 2030.
2. Reduce the number of people affected by disasters by 2030.
3. Reduce economic loss as measured by GDP by 2030.
4. Reduce disaster damage to critical infrastructure by 2030.
5. Increase the number of countries with disaster reduction strategies by 2030.
6. Enhance international cooperation to assist developing countries by 2030.
7. Increase the availability of multi-hazard early warning systems by 2030.

The implementation of this strategy for disaster disk reduction is guided by 13 **guiding principles** and four **priorities**.

14.4.3 Is there a better way of addressing global crises?

Several global crises face the world today. Each is addressed by its own strategy and action plan. It is natural to ask whether it is possible to address them all together—i.e. develop a strategy that applies to all crises.

The present strategies for global crisis management are at least in part due to the individual histories of their development process. The need to address climate change was first formally considered at COP1 in Berlin, Germany in 1995, and it took twenty years for the deliberations within the UN agencies to produce the Paris

Agreement of 2015. The SDGs are a sequel to the MGDs which were initiated in 2000. They were first suggested at the UN conference in Rio de Janeiro in 2012, three years before the term of the MGDs was due to end. It is therefore no surprise that these global strategies were developed along their own individual pathways, with goals that were firmly fixed on satisfying their own specific needs.

But one is still free to ask whether it is not possible to develop a common strategy that treats several issues together within the same framework. The advantages of such an approach are many. Obvious benefits include the favourable economics of such a choice. But perhaps more importantly, since all such actions are associated with the same cause (i.e. the betterment of human well-being), there will almost certainly be advantages in formulating a common conceptual framework for them all. The causal links (i.e. the reason why we are doing what we are doing) will be more obvious, providing more motivation for action. Such an integrated approach is the subject of the next section.

14.5 An integrated approach to the global challenges

14.5.1 How can we find better ways of solving our crises?

One of the possible ways of finding an integrated approach to addressing separate challenges is to combine the policy frameworks (that may have evolved separately) into a single framework. Finding such a strategy will require, amongst other things, a change in perspective.

To proceed, if we consider the climate change and sustainable development challenges in the first instance, an integrated solution can be found by combining the functional roles of the Paris Agreement with the SDGs into a common role. The problem then translates into finding a policy framework for such a common role.

As the Sendai Framework deals with a host of separate disasters, it may be instructive to consider it as a starting point for modelling the new framework. But a cursory examination soon reveals that this framework involves a range of measures from diverse sectors such as economic, legal, health and educational. This is both a **piecemeal** and **ad hoc** approach, and does not seem to possess an underlying rationale that permeates the whole framework. This is no doubt an effective framework in its own right. However, it lacks the elegance and simplicity that we desire. Examination of the SDGs reveals a similar a weakness.

What is needed is a first principles development of the framework. Such an ***ab initio*** development is more likely to reveal the reasoning behind the development, and will, in all probability, provide a more comprehensible and effective tool for action.

14.5.2 A suggested methodology

A possible methodology for the development of an integrated framework begins by adopting the broadest perspective to the issue. It then finds a unique concept that represents the situation at hand, and is able to provide clear linkages between the actions and the goals.

To illustrate, consider the problem of finding a common (integrated) framework for the challenges of climate change and sustainable development. The challenge being addressed in the former is how to contain climate change to a degree where its impacts are tolerable by humans. In the case of sustainable development, the challenge is to find satisfactory remedies to a host of issues that affect humans. It is interesting to note in passing that while the policy framework comprising the Paris Agreement is well-structured (according to the requirements stipulated in the first paragraph), the SDGs are not.

A common concept that reflects the essence of both challenges is **human well-being**. To proceed further, the individual concerns of both the challenges being considered in this unification process must be aligned with the properties of the chosen unique concept.

In summary, the possible methodology being proposed consists of three steps:

Step 1—find a common perspective.

Step 2—represent it with a unique concept.

Step 3—construct an integrated development framework that equates each issue that needs to be addressed by a property of the central concept.

(See research question 14.2—In search of an integrated development plan.)

Exercises and questions

Exercises

Exercise 14.1. Classifying human crises and disasters.

Two global crises are Climate Change and the El Nino-Southern Oscillation Cycle (ENSO) comprising the El Nino and La Nina weather phenomena.

(a) Classify each as either human-made or natural phenomena.

(b) List at least four impacts of climate change, classifying each as either physical or biological. Your list should contain at least two biological disasters.

(c) Draughts and flooding may be associated with both climate change and ENSO. Explain how each of these global crises influences the occurrence of draughts.

(d) Describe the conditions under which an impact of climate change can induce tsunami-like phenomena.

Questions

Question 14.1. Energy and well-being: a personal perspective.

(a) Assess your own economic status (according to your family income level) and consider the goods/services that you use in your own household. These will include, (amongst others) cooking, heating and cooling appliances, transportation, entertainment.

(b) How are the goods/services related to energy use? What are the forms of energy (traditional vs modern) that you use? Are there different types of energy used for the same purpose (e.g. cooking)?

(c) Next compare this with someone in your country whose economic status is very different from yours, and comment on the availability of these services to this other person (would they be able to obtain these goods and services on their family income?)

(d) How does the availability of different types of energy impact on the quality of your life?

Research questions

Research question 14.1. Traditional biomass and the health and well-being of women and children in poor countries.

Poor communities are dependent on traditional biomass for their cooking fuel, and traditional wood stoves for their cooking needs, because they cannot afford anything better.

Prepare a report on the use of traditional biomass and cooking stoves by communities in the Least Developed Countries of the world, with specific focus on

(i) the efficiency of traditional three stone stoves and the products of combustion;

(ii) the health effects of particulate matter and toxic gases produced by traditional stoves;

(iii) the economic impact on families of collecting firewood for their daily cooking needs.

Your report must explicitly discuss

- classification of particulate matter contained in the atmospheric emissions of traditional wood stoves into PM10, PM2.5 and PM0.1, and an assessment of their relative health impacts;
- methods by which the efficiencies of traditional stoves could be improved, including the use of 'rocket stoves';
- the environmental impacts, including deforestation, arising from the use of inefficient wood stoves.

Research question 14.2. In search of an integrated development plan.

Examine the Sendai Framework for Disaster Risk Reduction (SFDRR).

(a) Provide a brief summary of the essential features of the framework.

(b) Use the methodology suggested in section 14.5.2 to suggest improvements to the structure of this framework.

(c) Suggest a unique concept that represents the challenge being addressed.

References

[1] Millennium summit, 6–8 September 2000, New York. UN [accessed 7 October 2022]. Available from https://un.org/en/conferences/environment/newyork2000

[2] International Energy Agency 2002 *World Energy Outlook 2002* (Paris: IEA), 530 Available from https://iea.blob.core.windows.net/assets/11bc7b29-db2a-46fb-a515-1f5e7d2c7dfd/WorldEnergyOutlook2002.pdf

[3] Sustainable Development Goals Factsheet. UN [accessed 8 October 2022]. Available from https://un.org/sustainabledevelopment/wp-content/uploads/2015/08/Factsheet_Summit.pdf

[4] Energy poverty in the EU. European Commission [accessed 8 October 2022]. Available from https://energy.ec.europa.eu/topics/markets-and-consumers/energy-consumer-rights/energy-poverty-eu_en

[5] The clean energy for all Europeans package. EUI Florence School of Regulation [accessed 8 October 2022]. Available from https://fsr.eui.eu/the-clean-energy-for-all-europeans-package/

[6] Recommendation (EU) 2020/1563-Energy Poverty. European Union European Regional Development Fund [accessed 8 October 2022]. Available from https://projects2014-2020.interregeurope.eu/powerty/news/news-article/11957/recommendation-eu-2020-1563-energy-poverty/

[7] Sustainable Development Goals. UN [accessed 8 October 2022]. Available from https://un.org/en/sustainable-development-goals

[8] Sustainable Development. UN [accessed 9 October 2022]. Available from https://sustainabledevelopment.un.org/?menu=1300

[9] SDG Indicators Database. UN [accessed 9 October 2022]. Available from https://unstats.un.org/sdgs/indicators/database/

[10] Particulate Matter (Basics). US EPA [accessed 10 October 2022]. Available from https://epa.gov/pm-pollution/particulate-matter-pm-basics

[11] Disasters, Climate and Crises. IFRC [accessed 13 October 2022]. Available from https://ifrc.org/our-work/disasters-climate-and-crises

[12] Sendai Framework for Disaster Risk Reduction 2015–30. UNDRR [accessed 13 October 2022]. Available from https://undrr.org/publication/sendai-framework-disaster-risk-reduction-2015-2030

[13] UN World Conference on Disaster Risk Reduction. UNISDR [accessed 15 October 2022]. Available from https://wcdrr.org/